KU-189-984

AFTER

....

A DOCTOR EXPLORES WHAT NEAR-DEATH

EXPERIENCES REVEAL ABOUT LIFE

AND BEYOND

....

Dr Bruce Greyson

PENGUIN BOOKS

TRANSWORLD PUBLISHERS
Penguin Random House, One Embassy Gardens,
8 Viaduct Gardens, London SW11 7BW
www.penguin.co.uk

Transworld is part of the Penguin Random House group of companies
whose addresses can be found at global.penguinrandomhouse.com

Penguin
Random House
UK

First published in Great Britain in 2021 by Bantam Press
an imprint of Transworld Publishers
Penguin paperback edition published 2022

A CIP catalogue record for this book
is available from the British Library.

ISBN
9780552176835

Typeset in 11.5/15.75 pt Adobe Garamond Pro by Jouve (UK), Milton Keynes
Printed and bound in Great Britain by Clays Ltd, Elcograf S.p.A.

The authorized representative in the EEA is Penguin Random House Ireland,
Morrison Chambers, 32 Nassau Street, Dublin D02 YH68.

Penguin Random House is committed to a sustainable
future for our business, our readers and our planet. This book
is made from Forest Stewardship Council® certified paper.

Dr Bruce Greyson is Professor Emeritus of Psychiatry and Neurobehavioral Sciences at the University of Virginia School of Medicine. He was a co-founder and President of the International Association for Near-Death Studies and Editor of the *Journal of Near-Death Studies*.

His research for the past four decades has focused on near-death experiences, particularly their after-effects and implications. His NDE Scale has been translated into twenty languages and used in hundreds of studies worldwide. Dr Greyson has published over one hundred scholarly articles about near-death experiences and gives regular addresses at international conferences on the subject. *After* is his first book for general readers.

For those who faced death and generously shared with me their most personal and profound experiences

Contents

AFTER

AFTER

Introduction

....

A Journey into Uncharted Territory

Fifty years ago, a woman who had just tried to kill herself
told me something that challenged what I thought I knew
about the mind and the brain, and about who we really are.

The forkful of spaghetti was almost to my mouth when
the pager on my belt went off, launching the fork out of
my hand. I had been concentrating on the emergency
psychiatry handbook propped open between my tray and
the napkin holder, so the sudden beeping startled me. The
fork clattered to my plate, splashing tomato sauce on the
open page. I reached down to shut the pager off and noticed
a blob of spaghetti sauce on my tie as well. Cursing under
my breath, I wiped the blob off and then dabbed at it with
a wet napkin, which made it less colorful but a bit larger.
Only a few months out of medical school, I was trying
desperately to look more professional than I felt.

I walked over to the phone on the cafeteria wall and dialed the number on my pager's display. There was a patient in the emergency room who had overdosed, and her roommate was waiting to speak with me. I didn't want to take the time to walk across the parking lot to the on-call room, where I had a change of clothes, so I retrieved the white lab coat from the back of my chair, buttoned it up to hide the stain on my tie, and went down to the ER.

The first thing I did was to read the nurse's intake note. Holly was a first-year student at the university whose roommate had brought her to the hospital and was waiting for me down the hall in the family lounge. The nurse's and intern's notes said Holly was stable but not awake, and that she was sleeping in Exam Room 4 with a 'sitter' watching her, a routine precaution for psychiatric patients in the ER. I found her lying on a gurney, wearing a hospital gown, with a tube in her arm and heart monitor leads running from her chest to a portable machine that had been wheeled up next to the gurney. Her tousled red hair splayed across the pillow, framing a pale, angular face with a slender nose and thin lips. Her eyes were closed and she didn't move when I entered the room. On the gurney shelf under her was a plastic bag with her clothes.

I placed a hand gently on Holly's forearm and called her name. She didn't respond. I turned to the sitter, an older African American man reading a magazine in a corner of the exam room, and asked if he'd seen Holly open her eyes or speak. He shook his head. 'She's been out the whole time,' he said.

I leaned closer to Holly to examine her. Her breathing

was slow but regular, and there was no odor of alcohol. I assumed she was sleeping off an overdose of some medication. The pulse at her wrist was beating at a normal rate, but skipped a beat every few seconds. I moved her arms to check for stiffness, hoping that might give me a clue as to what drugs she had taken. Her arms were loose and relaxed, and she didn't wake up when I moved them.

I thanked the sitter and made my way to the family lounge at the far end of the hallway. Unlike the exam rooms, the family lounge had comfortable chairs and a couch. There was a coffee urn, and paper cups, sugar, and creamer on an end table. Holly's roommate, Susan, was pacing the room when I walked in. She was a tall girl with an athletic build, her brown hair pulled back tightly into a ponytail. I introduced myself and invited her to sit. Her eyes darted around the room, and then she sat down on one end of the couch, fidgeting with the ring on her index finger. I pulled up a chair next to her. The windowless room was not air-conditioned, and I was already starting to sweat in the heat of a late Virginia summer. I moved the standing fan a little closer and unbuttoned my white coat.

'You did the right thing, Susan, by bringing Holly in to the ER,' I started. 'Can you tell me what happened this evening?'

'I came home from a late afternoon class,' she said, 'and found Holly passed out on her bed. I called out and shook her, but couldn't wake her up. So I called the dorm counselor and she called the rescue squad to bring her here. I followed in my car.'

Still assuming Holly had overdosed on some medication, I asked, 'Do you know what drugs she had taken?'

Susan shook her head. 'I didn't see any pill bottles,' she said, 'but I didn't look around for any.'

'Do you know whether she was taking any medication on a regular basis?'

'Yeah, she was taking an antidepressant that she had gotten from the student health clinic.'

'Are there any other meds in the dorm that she might have taken?'

'I have some medication for my seizures that I keep in the cabinet in the bathroom, but I don't know that she took any.'

'Did she drink regularly or use other drugs?'

Susan shook her head again. 'Not that I've seen.'

'Does she have any other medical problems?'

'I don't think so, but I don't really know her that well. I didn't know her before we moved into the dorm a month ago.'

'But she was seeing someone at Student Health for depression? Had she been looking more depressed or anxious lately, or acting strangely?'

Susan shrugged. 'We weren't really that close. I didn't notice anything wrong.'

'I understand. Do you happen to know about any particular stresses she's been under lately?'

'As far as I know, she's been doing well in her classes. I mean, it's an adjustment for all of us starting college, being away from home for the first time.' Susan hesitated, then added, 'But she was having problems with this guy she was dating.' She paused again. 'I think he might have been pushing her to do things.'

'Pushing her to do things?'

Susan shrugged. 'I don't know. That's just the feeling I got.'

I waited for her to continue, but she didn't.

'You've been very helpful, Susan,' I said. 'Is there anything else that you think we should know?'

Susan shrugged again. I waited again for her to say something else, but she didn't. I thought I might have seen a slight shudder.

'How are you doing with all this?' I asked, touching her gently on the arm.

'I'm okay,' she said, too quickly. 'But I have to get back to the dorm. I've got a paper to write.'

I nodded. 'Well, thanks for bringing Holly in and for waiting to talk with me. Why don't you go back now and get to that paper? You can check on her in the morning if you want. We'll call you if we think of anything else.'

Susan nodded and stood up, and I walked her to the door. As I reached out to shake her hand, I again caught a glimpse of the stain on my tie and rebuttoned my lab coat so the ER staff wouldn't notice.

I walked back down the hallway to Holly's room to see if she'd awakened by then. She was still out cold, and the sitter confirmed that she hadn't stirred since I'd left. There wasn't much more for me to do that evening. I spoke with the medical intern evaluating Holly, who said that he was going to admit her to the intensive care unit to monitor her irregular heartbeat. I then called the faculty psychiatrist who was backing me up that night. He agreed there was nothing more for me to do at that point, but told me to make sure I documented everything and that I should check back on Holly and talk with her first thing in the morning. I would have to present her case to the senior psychiatrists on the consultation team on their morning rounds at eight a.m. As I walked across the parking lot to the on-call room, I congratulated myself on not making a fool of myself, and on my good fortune to have the patient admitted to the ICU so the medical intern would be responsible for her admission note and orders that night, rather than me.

When I entered the intensive care unit early the next morning, refreshed with a good night's sleep and a change of clothes, I scanned the rack at the nursing station for Holly's medical chart. One of the nurses was writing in it, and looked up at me.

'You're from Psychiatry?' she asked.

I nodded and said, 'I'm Dr Greyson.' It was not hard to identify me as the shrink, as I was the only one in the ICU wearing street clothes under my white coat, rather than scrubs.

'Holly's awake now, and you can talk with her, but she's still pretty drowsy,' the nurse said. 'She's been stable all night except for a few PVCs [premature ventricular contractions].' I knew that those irregular heartbeats could mean nothing, but they could also be related to whatever medications she had taken the night before.

'Thanks,' I said. 'I'll go speak with her briefly now, but the consultation team will be here in about an hour to interview her. Do you think she'll be stable enough to be transferred to the psychiatry unit today?'

'Oh, yeah,' the nurse said, rolling her eyes. 'There are patients stacked up in the ER waiting for a bed here to open up.'

I walked over to Holly's room and knocked on the jamb of the open door. She now had a tube in her nose as well as in her arm, and the heart monitor leads were now connected to a screen above her bed. I pulled closed the curtain around her bed behind me, and softly called her name. She opened one eye and nodded.

'Holly, I'm Dr Greyson,' I said. 'I'm with the psychiatry team.'

She closed her eye and nodded again. After a few seconds, she mumbled softly, her speech a bit slurred, 'I know who you are. I remember you from last night.'

I paused, replaying in my mind our encounter the night before. 'You looked like you were asleep in the ER last night,' I said. 'I didn't think you could see me.'

Her eyes still closed, she muttered softly, 'Not in my room. I saw you talking with Susan, sitting on the couch.'

That caught me up short. There was no way she could have seen or heard us talking at the far end of the corridor. I wondered whether that wasn't her first visit to the ER, and whether she could have guessed that I'd talked with Susan there.

'The staff told you that I spoke with Susan last night?' I suggested.

'No,' she said, more clearly now. 'I *saw* you.'

I hesitated, not sure how to proceed. I was supposed to be leading this interview, gathering information about her thoughts of harming herself and what was going on in her life. But I was confused, and didn't know how to proceed. I wondered whether she was just toying with me, the new intern, trying to rattle me. If so, she was doing a good job. She sensed my uncertainty and opened both eyes, making eye contact for the first time.

'You were wearing a striped tie that had a red stain on it,' she said firmly.

I leaned forward, very slowly, wondering whether I'd heard her correctly. 'What?' I said, barely able to form the word.

'You were wearing a striped tie with a red stain on it,' she repeated, glaring at me. She then went on to repeat the conversation I'd had with Susan, all my questions and Susan's answers, along with Susan's pacing and my moving the fan, without making any mistakes.

The hair rose on the back of my neck and I felt goose bumps. She couldn't possibly have known all that. She might have guessed what questions I'd be likely to ask, but how could she have known all the details? Had someone already spoken with her earlier that morning, and told her what I'd written in my note? But no one else had been in the room with Susan and me. How would anyone else know the details of what we had said and done? And no one outside the family lounge had seen the stain on my tie the night before. There was no way Holly could have known that I had spoken with Susan, let alone been familiar with the content of our conversation or the stain on my tie. And yet she did. Every time I tried to focus on what she'd said, I found my thoughts getting muddled. I couldn't deny that she knew the details of my conversation with her roommate. I'd heard it with my own ears; it definitely happened. But I couldn't figure out *how* she knew them. I told myself that it had to be a lucky guess or some kind of trick.

But I couldn't fathom how such a trick could have been pulled off. Holly was just waking up from her overdose. She hadn't spoken with her roommate since the day before. How could she know what Susan and I had said? Could Holly and Susan possibly have colluded before her overdose, planning what Susan would say to me? But they couldn't have colluded to drop spaghetti sauce on my tie. Besides, Susan had been agitated when I spoke with her in the ER, and Holly was now still groggy and depressed. It didn't look or feel like a hoax.

I had no answers to these questions, but I also had no time to think about them, and I had no convenient box to put them in. This was years before anyone in the English-speaking world had heard the term 'near-death experience.' I was stymied by this incident because I couldn't explain it. All I could do was file these questions away in the back of my mind.

Holly's erratic breathing, indicating that she had fallen back asleep, pulled me back to the present. *My* bewilderment could not be the issue that day. My job was to help Holly deal with *her* issues and help her resolve her problems and find some reasons to live. For now, I had to focus on learning what I could about her life stressors and assessing her suicidal thoughts before the team made rounds.

I touched her arm gently and called her name again. She opened one eye, and I tried to continue my interview. 'Holly, can you tell me about your overdose last night and what led up to it?' I pulled it together enough to get from her that she had taken an overdose of Elavil, which can cause dangerous heart rhythms, and that she had taken 'a few' previous overdoses in high school. She corroborated everything Susan had told me, adding a few additional details. She told me she was feeling overwhelmed by the social pressures of college and felt she didn't fit in with her peers. She said she wanted to drop out of the university and return home and go to a local community college, but her parents kept telling her to give it more time. When she seemed to be falling asleep again, I thanked her for speaking with me and told her the psychiatry team would be coming around to see her in about an hour. She nodded and closed her eyes.

I put in a call to the student health clinic and left a message that Holly had been admitted, and to request the records of her psychiatric treatment there. I then wrote up a brief intake note, based largely on what Susan had told me the night before and on what little I had observed about Holly's mood and thought processes that morning. But my presentation to the psychiatry consultation team was hardly complete. I pointedly avoided any mention of her claim to have seen and heard me while she was asleep in a different room, and decided then and there not to let any of my colleagues know about it, at least until I could come up with a reasonable explanation. At best, they would think I'd lost it and was acting unprofessionally. At worst, they might wonder if I had *really* lost it and was imagining the whole thing.

It was clearly impossible, I told myself, for Holly to

have seen or heard what was happening in the family lounge while she had been asleep at the far end of the emergency room. There had to be some other way she'd learned about it. I just couldn't figure out what that other way might be. None of the nurses in the ICU knew about my conversation with Susan in the ER, nor did any of the ER staff on duty the night before know the details that Holly had shared. As unsettling as this incident was for me, a green intern trying to feel that I knew what I was doing, I could only tuck it away, with uncertain plans to return to it sometime in the future. I didn't even tell my wife, Jenny. It was just too weird. I would have been embarrassed to tell someone this had happened, and that I was taking it seriously. And I also knew that telling someone would make it harder to lock it away, and I'd be forced to deal with it somehow.

I believed that there must be some reasonable physical explanation for how Holly knew these things, and I'd have to find that explanation myself. And if there wasn't . . . well, the only alternative was that the part of Holly that thinks and sees and hears and remembers somehow left her body and followed me down the hall to the family lounge and, without the benefit of eyes or ears, took in my conversation with Susan. That made no sense to me at all. I couldn't even imagine what it would mean to leave my body. As far as I could tell, I *was* my body. But I couldn't afford to think about these things at this point in my life. I was not in a position to investigate the incident, to track down Susan and ask if she'd noticed the stain on my tie and, if so, whether she'd mentioned it to anyone, and to track down the nurses who'd been working in the ER the night before – not to mention tracking down anyone who might have seen me drop my fork in the cafeteria and then talk with Holly, as unlikely as that would have been. Nor was I in the mindset to investigate the incident. I just wanted it to go away.

For the past half century, I've been trying to understand how Holly could have known about that spaghetti stain. Nothing in my background or scientific training to that point had prepared me to deal with such a frontal assault on my worldview. I had been raised by a no-nonsense skeptical father, for whom life was chemistry, and I followed his lead in forging my own career as a mainstream scientist. As an academic psychiatrist, I have published more than a hundred scholarly articles in peer-reviewed medical journals. I have been fortunate enough to serve on the full-time medical school faculty at the University of Michigan, where I ran the emergency psychiatry service; at the University of Connecticut, where I was clinical chief of psychiatry; and at the University of Virginia, where I held the endowed Chester F. Carlson Professorship of Psychiatry and Neurobehavioral Sciences. Being in the right place at the right time has enabled me to receive research grants from government agencies, from pharmaceutical companies, and from private, nonprofit research foundations. I have been privileged to serve on grant review boards and program planning workshops at the National Institutes of Health, and have addressed a symposium on consciousness at the

United Nations. I have earned awards for my medical research and been elected a Distinguished Life Fellow of the American Psychiatric Association.

Overall, I have had a very satisfying career as an academic psychiatrist – thanks in large part to the brilliant and supporting mentors and colleagues who deserve a lot of the credit for my success. But through all those years, in the back of my mind were the nagging questions about the mind and the brain that Holly raised with her knowledge of that stain on my tie. My personal need as a skeptic to follow the evidence kept me from closing my eyes to events like that – events that seemed impossible – and led me on a journey to study them scientifically.

I'd become the director of the psychiatric emergency service at the University of Virginia when Raymond Moody began his training there in 1976. When Raymond's book *Life After Life*, the first book in English to use the term 'near-death experience' and the acronym NDE, became a surprise bestseller, he was quickly inundated with letters from readers who'd had such experiences. As an intern without the time to respond to all those letters, he turned to me, as his training supervisor in the emergency room, for help. And I was stunned to learn at that time that Holly's experience, which had knocked me for a loop, was not at all unique. Raymond had interviewed other patients who claimed to have left their bodies and observed what was going on elsewhere, while they were close to death.

That revelation grabbed my attention, and launched me

on a journey to follow an evidence-based approach to NDEs. If I hadn't met Raymond and read his groundbreaking book, I might never have followed the trail of that spaghetti stain. But I soon learned that NDEs were not a new phenomenon. I discovered a multitude of NDE accounts from ancient Greek and Roman sources, all the major religious traditions, narratives collected from indigenous populations around the world, and the medical literature of the nineteenth and early twentieth centuries.

With colleagues at other universities who had also stumbled across NDEs, I cofounded the International Association for Near-Death Studies (IANDS), which would serve as an organization to support and promote research into these experiences. For more than a quarter century, I served as the director of research for IANDS and edited the *Journal of Near-Death Studies*, the only scholarly journal dedicated to NDE research. Over the decades I assembled a collection of more than a thousand experiencers who were kind enough to fill out questionnaire after questionnaire for me, some for more than forty years. I was able to compare the findings from those 'volunteers' with the NDEs of patients hospitalized, for example, for cardiac arrest, for seizures, and for attempted suicide. And along that journey, I discovered some common and universal themes in these experiences that go beyond cultural interpretations, as well as patterns of consistent aftereffects on individuals' attitudes, beliefs, values, and personalities. And I have been able to show that these experiences can't be dismissed simply as dream states or hallucinations.

What I found in that forty-five-year journey was a

record of NDEs that goes back centuries and encircles the globe. I discovered that NDEs are common, and play no favorites. Even neuroscientists have them. When neurosurgeon Eben Alexander was stricken by a rare brain infection that plunged him into a weeklong coma, from which he awakened with vivid memories of an elaborate near-death experience, he came to my office to help make some sense of this seeming impossibility.

I discovered, over almost a half century of struggling to understand near-death experiences, that their impact extends far beyond the individual experiencer. The more I learned about them, the more they seemed to cry out for an explanation beyond the limited understanding of our everyday ideas about the mind and the brain. And those new ways of thinking about our minds and our brains open up the possibility of exploring whether our consciousness might continue after the death of our bodies. And that, in turn, challenges our concept of who we are, how we fit into the universe, and how we might want to conduct our lives.

Some of my scientist colleagues have warned me that my open-minded approach to exploring 'impossible' experiences like NDEs would open the floodgates to all sorts of superstitions. As a skeptic, I say bring them on! Let's not prejudge them because of our beliefs; let's test those challenging ideas and see whether they are in fact superstitions – or whether they're windows into a more comprehensive picture of the world. Far from leading us away from science and into superstition, NDE research actually shows that by applying the methods of science to the nonphysical aspects of our world, we can describe reality much more accurately than if we limit our science to nothing but physical matter and energy.

In following the scientific evidence that has accumulated over the past several decades, and not promoting any one theory or belief system, I know I will disappoint many of my friends who may favor one or another particular view. I know that some of my spiritual friends may object that I take seriously the possibility that NDEs may be brought on by physical changes in the brain. And I know that some of my materialistic friends may be dismayed that I take seriously the possibility that the mind may be able to function independent of the brain. And I know that some in both camps may complain that by not taking sides, I am taking the easy way out.

But in fact, intellectual honesty demands that I avoid taking sides in this debate. I think there is enough evidence to take seriously *both* a physiological mechanism for NDEs *and* continued functioning of the mind independent of the brain. The belief that NDEs are due to an unidentified physiological process is plausible, and consistent with the philosophical view that the real world is purely physical. On the other hand, the belief that NDEs are a spiritual gift is also plausible, and consistent with the philosophical view that there is a nonphysical aspect to who we are. But neither of these ideas, while plausible, is a scientific premise – because there is no evidence that could ever disprove either of them. They are instead articles of belief.

As I hope to show in this book, there is no reason NDEs can't be *both* spiritual gifts *and* enabled by specific physiological events. The scientific evidence suggests that both ideas can be true without any conflict – which allows us to move beyond the artificial divide between science and spirituality. But my openness to both views doesn't mean that I have no opinions about the meaning of near-death experiences.

The decades of research have convinced me that near-death experiences are quite real and quite profound in their impact, and are in fact important sources of spiritual growth and insight – whatever their source. I know that they matter critically to the experiencers themselves in the way they transform their lives. I believe that they also matter to scientists in that they hold vital clues to our understanding of mind and brain. And I think they also matter to all of us in what they tell us about death and dying, and more important, about life and living.

Throughout the body of this book, I've skipped over the methodological and statistical details of my research, but those who want the technical details of the studies I mention will find them described in the references cited in the notes in the back of the book. All of my peer-reviewed journal articles can be downloaded from the University of Virginia Division of Perceptual Studies website at www.uvadops.org.

Although this book is based on my forty-five years of scientific research into NDEs, it was not written specifically for other scientists. And although I hope people who have had NDEs will feel that I have done justice to their experiences, I have not written this book specifically for them. Rather, I've written this book for the rest of us, for those who are curious about the incredible scope of the human mind and about the deeper questions about life and death.

A lot has been said and written about dying and what may come after – much of it pitting scientific and religious viewpoints against each other. I try in this book to move that discussion forward and help change the dialogue. I hope to show that science and spirituality are compatible, that being spiritual doesn't require you to abandon science. This journey has taught me that approaching the world scientifically, basing our beliefs and understanding on evidence, doesn't have to stop us from appreciating the spiritual and nonphysical aspects of our lives. And on the other hand, appreciating the spiritual and nonphysical doesn't have to stop us from evaluating our experiences scientifically, basing our beliefs and understanding on the evidence. Though I learned a lot about dying and what might come after, this is not a book solely about death. It is also a book about life and living, about the value of compassion and our interconnectedness with one another, and about what makes a life meaningful and fulfilling.

My aim in writing this book is not to convince you to believe any one point of view, but to make you think. I hope to show that a scientific perspective can help us understand what NDEs tell us about life and death, and about what may come after. By following the scientific evidence, I've learned a lot about near-death experiences and what they mean. I wrote this book to share with you my passion for that journey. My goal is to make you think about the questions and ponder the answers – not to make you believe any one point of view, but rather reevaluate how you think about life and death. I'm not a Moses, handing down the Ten Commandments. I'm a scientist sharing what I think the data suggest.

As desperately as I wanted to erase from my memory my entire encounter with Holly, I was by then enough of a scientist to know I couldn't just ignore it. Pretending something didn't happen just because we can't explain it is the exact opposite of science. My quest to find a logical explanation for the riddle of the spaghetti stain led me into a half century of research. It didn't answer all my questions, but it did lead me to question some of my answers. And it would soon take me into territory I never could have imagined.

A Science of the Unexplained

I had never met someone with half a face. Six months into my psychiatric training, Henry was admitted to my hospital. When I first saw him lying on his hospital bed, it was hard not to stare at the right side of his face, where his jaw and his cheek should have been. The plastic surgeons had done a remarkable job of piecing together skin grafts from his belly to close up the wounds on his face, but even so, I had a hard time maintaining my composure when I looked at him. He spoke in slow and slightly slurred words, using only the left side of his mouth. But as awkward as I felt, he didn't seem at all embarrassed or reluctant to talk with me. In fact, he seemed calm and composed when he told me what had happened after he shot himself.

Then in his forties, Henry had been the youngest child in a poor farming family. His older siblings had all moved away from the family farm when they married, but Henry, despite marrying, never left home. When he was twenty-three, his father suffered a heart attack while he and Henry were hunting. Henry managed to carry his father back to the farm, only to watch his father die in his arms. His mother then took over responsibility for managing the farm, and a few years later, Henry's wife left him, taking their children to live with her parents in town.

Ten months before he shot himself, Henry's mother came down with pneumonia and he drove her to the hospital, where she was admitted. She asked him not to leave her side, but he went home that night to tend to the chickens. When he returned the next morning, she was unconscious. She died a few hours later.

Henry was devastated and began drinking heavily. Racked with guilt about having abandoned her in the hospital, he had nightly dreams of his mother being alive. He could not bring himself to touch any of her personal effects, and had left everything in the house just as she'd left it. When he drank, he would become despondent, muttering over and over that 'home just isn't home anymore.' Finally, after several months of depression, and after spending the entire morning drinking, he went to the cemetery where his parents were buried, taking his hunting rifle.

After sitting on the grave for a couple of hours, reliving and imagining conversations with them, he decided it was time to join them. He lay down on the grave, positioning his head over where he thought his mother's breast would be. Henry lodged the .22-caliber hunting rifle between his legs, aimed it up at his chin, and gently squeezed the trigger with his thumb. The bullet tore through the right side of his face, leaving a trail of shell fragments embedded in his cheek and temple, but by a stroke of luck the bullet missed his brain.

I tried to keep my voice steady and avoid staring at his stitched-up cheek as I interviewed him. 'That sounds pretty painful,' I suggested. 'I can only imagine what must have been going through your mind. What was it like for you?'

The left side of Henry's face curled into a half smile. 'As soon as I squeezed the trigger,' he said, 'everything around me disappeared: the rolling hills, the mountains behind them, all vanished.'

He looked up at me and I nodded and asked, 'And what then?'

'I found myself in a lush meadow of wildflowers. There, welcoming me with open arms, were my mama and papa. I heard Mama say to Papa, "Here comes Henry." She sounded so happy to see me. But then she looked right at me and her expression changed. She shook her head and said, "Oh, Henry, now look what you've done!"'

Henry paused, looked down at his hands, and swallowed. I waited a moment, and then said, 'That must have been hard for you. What did that feel like?'

He just shrugged and shook his head, then took a deep breath. 'That was it,' he said. 'Then I was back in the cemetery, and they were gone. I felt the warm puddle of blood under my head and thought I'd better get help. I started to drag myself toward my truck, but before I got there, a gravedigger saw me and ran over. He wrapped a piece of cloth around my head and drove me to the hospital.' He shrugged again. 'And here I am.'

'That's quite an experience,' I offered. 'Had you ever seen your parents before, since they'd passed on?'

He shook his head. 'Nah. But it sure felt good to see them together there.'

'It sounds like you must have blacked out, at least briefly, after you shot yourself. Do you think seeing your parents might have been a dream?'

Henry pursed his lips and shook his head. 'That was no dream,' he said. 'Meeting Mama and Papa was every bit as real as meeting you right now.'

I had to pause there to try to make sense out of what he was saying. It made perfect sense to Henry – he saw them because they were welcoming him into heaven. But in my scientific worldview, that kind of thing couldn't be real. I ran through the possibilities in my mind. Was Henry psychotic? Had he been so drunk that he'd been hallucinating? Had he spent so much time sitting on his parents' grave that he was in alcohol withdrawal and having delirium tremens? Was that vision of his parents just a part of his grieving?

I couldn't make a case for Henry being crazy. At this point, after several days in the hospital, he was speaking calmly and there was nothing strange about the way he was acting. He hadn't had any physical signs of alcohol withdrawal since he'd been in the hospital. And to my surprise, he didn't seem at all sad.

'When you pulled the trigger,' I asked Henry, 'what were you hoping would happen?'

'I just didn't want to live anymore,' he said quickly. 'I

didn't care what happened. I'd just had enough and couldn't go on without Mama.'

'And now? What are your thoughts about ending it all now?'

'I don't think about that at all now,' he said. 'I still miss Mama, but I'm happy now that I know where she is.'

In my short time as a psychiatrist in training, I'd never seen someone who'd survived a suicide attempt and come out of it seeming as confident as Henry did. He said he was ashamed of his suicide attempt, but grateful for his vision. And he was eager to talk with other patients, to reassure them about the value and sanctity of life. Whatever had led him to see his parents, this vision was clearly helping him cope with his grief.

This was still several years before the term 'near-death experience' was introduced into the English language, and the only framework I had for understanding Henry's experience was that of a hallucination, an imaginary reunion with his deceased parents. I viewed his experience as a psychological defense mechanism and nothing more.

This was just a few months after Holly had told me about seeing the stain on my tie, and I was still trying to make sense of that incident. But Henry's experience felt very different to me than Holly's did. She claimed to have seen and heard things far away from her unconscious body, but still within the normal physical world. She did not report seeing or hearing any spirits. Henry, on the other hand, claimed to have seen and heard the spirits of his deceased parents. But the biggest difference was that I could look at Henry's vision from an objective scientific viewpoint. Holly, on the other hand, had dragged me personally into her vision, throwing me off-balance whenever I tried to contemplate it, and leaving me grasping futilely for explanations.

I could label Henry's vision as a psychological defense mechanism. But how could I possibly convince him that it hadn't been real? I knew that if I told him he'd imagined the whole thing, I'd lose any rapport I had with him as his doctor. I could also see how helpful that vision was for him, and how important it was to resolving his suicidal thoughts. I viewed his vision as a hallucination that his unconscious mind had created to help him cope with the death of his mother. I decided that the way I could be most helpful to Henry, as his doctor, was to reinforce the value of his vision, and not to challenge the one thing that was giving him a reason to live. My message to him was straightforward: 'It sounds like you've had a very powerful experience that's given you a new purpose in life. Let's look at what it means to you and where you go from here.'

My intent was to explore with Henry the symbolic meaning of his vision as a way of reuniting psychologically with his deceased mother, but he took his visit with his parents concretely and not as a symbol of anything. It never occurred to me at the time that he might have viewed that visit as real simply because it *was* real. Nothing in my background or training to that point suggested that Henry really could have seen his parents. I'd been raised by a chemist whose perception of reality was defined by the periodic table of the elements.

*

My father was a chemist by day, and by night . . . well, he was a chemist then, too. He built a chemistry lab in the basement of every home we lived in during my childhood. Second only to his passion for science was his joy at sharing it with others. While I was still in elementary school in Huntington, New York, he taught me to use a Bunsen burner, a balance beam scale, a centrifuge, a magnetic mixer, a graduated cylinder, and Erlenmeyer and round-bottom flasks.

Many of my father's experiments involved Teflon in the early days after it was discovered by accident by a scientist at DuPont. My father worked at a small chemical company that made things out of Teflon, such as wiring insulation and rocket fuel cells. The main advantage of Teflon over other coatings was that its surface was so slippery almost nothing stuck to it. Some of my father's creations led to useful advances. He sprayed my mother's pots and pans and spatulas with various forms of Teflon years before Teflon-coated cookware became commercially available – although from time to time we'd find bits of it in our food. Other inventions of his were less successful. He put Teflon inserts into our shoes to prevent them from giving us blisters. They were so slippery that with each step, my foot slid around inside my shoe. Walking became tricky, and running was downright dangerous. Whether or not his experiments worked out was less important to my father than the excitement of doing them, the uncertainty of whether or not they would pan out.

A quiver of anticipation ran down my spine as I lay faceup on the sacrificial stone. Sunlight filtered through the towering pines, highlighting the mountain laurel and rhododendron bushes, and birds twittered in the morning breeze. There was a groove a half inch deep in the surface of the large granite slab, completely surrounding my body, and just below my feet there was a short gutter cut between the circular groove and the edge of the slab. The entire slab, which must have weighed more than a ton, rested a few feet off the ground, atop four stone bases.

My father, a short, broad-shouldered man with a twinkle in his eye, paced around the slab with a tape measure in his hand and a pipe in his mouth, writing notes and drawing diagrams in his notebook. The dozen or so stone chambers, walls, and drains surrounding the granite slab, and the upright stones that seemed to line up with views of the sun at certain times of the year, were a mystery. In fact, the farmer who owned the land, in Salem, New Hampshire, in the mid-twentieth century called it 'Mystery Hill.' Others who had studied this site speculated that it might have been built by Viking settlers around AD 1000, hundreds of years before Columbus came to America, or by Celts from the British Isles around 700 BC, or by various Abenaki and Pennacook Indian tribes over thousands of years.

Whatever its origin, lying on that cold slab sent shivers down my spine. I could imagine my blood collecting in the groove surrounding my body, to be channeled down the gutter below my feet into a collecting bucket. It was creepy, but it was also exciting. There I was, a ten-year-old boy, helping my father try to solve a scientific mystery. I couldn't tell whether my shaking was more a reaction to the cold of the stone slab in the crisp New England fall, or to the thrill of discovery. For my father, it was obviously the latter, and I was already catching his excitement at taking part in the March of Science, pushing back the borders of the Unknown. At ten, I was already hooked on science, on answering questions by collecting and analyzing data rather than by armchair speculation or by taking rumors and folktales at face value.

The truth about Mystery Hill remains fuzzy to this day, probably because multiple groups of people altered the ruins over the centuries, destroying or changing the evidence of its origin. The 'sacrificial stone' may be just the bottom half of a nineteenth-century cider press, with the gutter around the edge to collect the liquid from the pulped apples, or a stone press to extract lye from wood ash for making soap. My father and I found nothing to support any of the claims about Mystery Hill, but I never forgot the excitement of a systematic quest to find the truth.

Always the skeptic, my father entertained continual doubts about his interpretation of things. He was happiest investigating things that he didn't understand, or things that contradicted his expectations. And he passed on to me not only his passion for science but also his awareness of the essential tentative nature of science. Science by its very nature is always a work in progress. No matter how well-founded we think our worldview is, we have to be prepared to rethink it if new evidence raises doubts. One of the fruits of that open-minded attitude is an appreciation for things we can't explain. Studying things that fit our preconceived ideas helps us understand their fine points better. But studying things that *don't* fit our preconceived ideas is what often drives breakthroughs in science.

Even though my father encouraged me to investigate things I couldn't explain, he never mentioned the mind, or abstract things like thoughts and feelings – let alone even more abstract concepts like God or spirit or soul. I felt quite fulfilled by my scientific upbringing and my plans for a scientific career, and following my father's lead, I embraced empirical evidence as my standard for finding the truth.

As an undergraduate student at Cornell University, I majored in experimental psychology, applying scientific methods to study how goldfish learned to find their way through a maze, how rats learned to press a bar for food at certain times but not others, and how immature rhesus monkeys learned to find food under one kind of object but not another. But as fascinated as I was by animals' intelligence, my desire to work with people led me from college into medical school. There were many things I enjoyed about med school, from delivering babies to making home visits to elderly patients. But the more I learned about mental illness, the more I appreciated how little we understood the brain, and the allure of the unanswered questions eventually began pulling me toward psychiatry.

On a visit to my parents' home during my third year of med school, I surprised my father with the news that I was thinking of becoming a psychiatrist. I told my father that I was fascinated by the effects our unconscious thoughts and feelings had on our behavior. Sitting in his easy chair with his legs crossed, my father slowly pulled a corncob pipe and tobacco pouch out of his jacket pocket. He meticulously filled the bowl of the pipe and tamped down the tobacco, then added some more and tamped it down again. Then he struck a wooden match and carefully waved it over the bowl as he drew gently on the pipestem. Finally, he looked up and, to my surprise, he asked, 'What makes you think we have unconscious thoughts and feelings?'

I was shocked by this blunt challenge. But my father wasn't saying that the unconscious *didn't* exist. He was just asking for the evidence – as any skeptical scientist should. Nevertheless, his question took me aback. The influence of our unconscious – things we think and feel without being aware of them – has been the bread and butter of psychiatry for at least a hundred years.

Sigmund Freud compared the mind to an iceberg. Thoughts and feelings we are aware of are like the tip that's visible above sea level. For example, you are aware that you feel thirsty and you consciously decide to get a drink. But nine-tenths of the iceberg, unseen beneath the ocean surface, is the unconscious, the thoughts and feelings that we're not aware of, but that influence our behavior anyway. For example, most classroom teachers would not knowingly give better grades to the most attractive students. But there is ample evidence that they do in fact give higher grades to more attractive students, without even being aware of it. The idea that our unconscious thoughts and feelings affect our behavior was one of the many things I accepted on faith – faith in my professors and in the textbooks – without questioning.

As surprised as I was that my father questioned the role

of unconscious thoughts and feelings, I realized that he had a point. I should look into the evidence of the unconscious before accepting it. But that raised the question of just what constitutes evidence when it comes to things you can't see and measure, like thoughts and feelings. Although scientists have made giant strides in understanding the physical part of our world, we also experience nonphysical things, such as thoughts and emotions. Those nonphysical things are just as much a part of our world as are physical objects like chairs and rocks. Scientists can make observations and collect data about *them*, just as we can about physical objects.

In fact, there is a long tradition of scientists studying phenomena that can't be observed directly, from emotions to subatomic particles. We can't directly observe emotions – such as love, anger, or fear. But we can study them *indirectly*, by looking at how they affect our words, behavior, and bodily reactions. For example, when we feel anger – a nonphysical emotion – our speech may get louder and more abrupt, our foreheads may wrinkle and our blood pressure rise, and we may slam things down on tables and counters. And from those observable effects, others can infer that we are angry.

Likewise, physicists can't directly observe some subatomic particles that are too small and too short-lived to capture. But physicist Donald Glaser won the 1960 Nobel Prize in Physics for studying them *indirectly*. He showed that by shooting tiny, short-lived particles through a bubble chamber – a vessel filled with a fluid like liquid hydrogen – we can study the trail of bubbles the particles leave in the liquid. And from those trails we can learn a lot about the particles themselves.

It was precisely this scientific tradition of following the evidence that showed me the limitations of the worldview I had been taught. There were lots of things that couldn't be fully explained in terms of physical particles and forces, but that happened anyway. It didn't seem scientific to shy away from some things just because they were hard to explain. Those things that didn't fit my worldview cried out to me to try to understand them, rather than write them off. Respecting things that are difficult to measure, rather than dismissing them as unreal, is not rejecting science. It's *embracing* science.

As a psychiatrist in training, I treated some hospitalized patients who believed they were reading other people's minds. I assumed, as did most psychiatrists, that these ideas were based on wishful thinking and confusing fantasy with reality. But did we have any evidence to back that up? How did we know that these patients' beliefs – that they could read minds – were a symptom of their mental illness, and not real? Of course, as a scientist, I couldn't just accept their claims as real without testing them. But I also couldn't just dismiss their claims as delusions, either, without looking into them. I thought that either validating their beliefs or rejecting them without evidence was doing these patients a disservice, as well as violating scientific principles. So I designed and, with my fellow trainees, carried out a controlled experiment to test whether these patients really *could* read minds.

I was a bit uneasy about the risks of carrying out such a study. As a scientist, I wanted to know whether or not these patients could provide evidence of their claims. But part of my job as a psychiatrist was to persuade delusional patients to give up their false beliefs and think more realistically. If my patients' beliefs in mind reading were unrealistic, would it only reinforce their false ideas by taking them seriously?

I wondered whether the potential benefits of this research would offset the potential risks to the patients themselves. I therefore discussed my proposed study with the medical and nursing staffs on the psychiatry ward. I acknowledged my hesitation about doing this kind of study, and my fear that if I treated these patients' unusual beliefs seriously, I might only solidify their delusions. But to my surprise, the ward director and the staff found the study intriguing, and felt that in the safe environment of the hospital, they'd be able to handle any worsening of the patients' symptoms, if there were any. With the blessing of the staff, then, I went ahead. Two of my fellow psychiatry trainees volunteered to alternate being the 'senders' in the experiment – that is, the people whose minds the patients would try to read.

The patients sat alone, one at a time, in a reclining chair in my office, relaxing for a few minutes. Then, when they felt ready, they spoke into a tape recorder, describing any images or impressions that came to them. Meanwhile, the 'sender,' in another office down the hall, concentrated on a randomly selected magazine picture showing a calm, scary, aggressive, funny, or erotic scene. After five minutes I entered my office and handed the patients an envelope with five magazine pictures. The patients then rated the five pictures as to how closely each one matched their impressions. When they finished, I told them which picture the 'sender' had been concentrating on, and we spent a few minutes discussing the session.

The study turned out just as my colleagues and I expected. None of the patients showed any evidence that they had read the 'sender's' mind. There was no indication that their belief in mind reading had any basis in reality. But there was another finding of the study that I hadn't anticipated. After it was over, I asked each of the patients how they felt about it. To my surprise, they were all happy that they had participated, and – more important – they all felt more trustful of the hospital staff because we had taken their thoughts and feelings seriously enough to test them. In addition, one of the patients added that failing to read the 'sender's' mind in this study led him to doubt his other irrational ideas, and helped him be able to separate fantasy from reality. His therapist independently told me that during the course of this experiment, the patient got markedly better. None of the patients reported any worsening of their illness as a result of the experiment.

Carrying out this experiment brought back the excitement I'd felt lying on that 'sacrificial' stone slab at Mystery Hill. I was collecting data to test an idea most of my colleagues wouldn't have bothered with, an idea my research might show to be wrong but that held the potential to change our thinking about mental illness. The fact that the patients couldn't read minds confirmed what we'd expected, but that wasn't what excited me. The thrill for me was in using science to test a provocative idea. The process was more important to me than the answer. My report of that experiment was later published in a mainstream medical journal, and it won the national William C. Menninger Award that year for the best research report by a trainee in neurology, psychiatry, or neurosurgery.

It wasn't until several years later that I met Raymond Moody and first heard about near-death experiences. Raymond began his psychiatric training at the University of Virginia the year I started teaching there as the newest addition to the psychiatric faculty. His first clinical rotation was in the emergency room, where I supervised all the trainees. I knew that Raymond had taught philosophy before he went to medical school, and that he had written a book while still a medical student, but I didn't know what his book was about. One day, during a quiet time in the ER, we got talking about his background and he told me about his book, called *Life After Life*, in which he used the term 'near-death experience' to label the unusual experiences some people had when they seemed to be on the threshold of death. As he talked, it gradually dawned on me that what he had described in his book was related to what had happened years earlier, both to Henry when he thought he saw his deceased parents, and to Holly when she saw me speaking with her roommate while her body lay unconscious in another room. Both Holly and Henry described at least some of the features Raymond had found in NDEs. I'll never know whether they experienced more NDE features because I hadn't known at the time to ask. But it was a revelation to learn that other doctors had heard of these experiences, and had even given them a name! It felt like a door was beginning to crack open.

I had come to the University of Virginia knowing about its Division of Perceptual Studies, a research unit founded by the late Ian Stevenson, the former chairman of psychiatry. For decades, Ian had been collecting and studying the same sort of unexplained experiences that Raymond had described in his book. Of course, Ian hadn't called them near-death experiences until Raymond introduced us to the term. He had filed them under various categories such as 'out-of-body experiences,' 'deathbed visions,' and 'apparitions.'

I took Raymond to meet Ian, and the three of us discussed how to study these experiences in a scientific way. Raymond was getting a huge volume of letters every week, and when I started reading them, I found they all had the same theme. Almost all of the letter writers had been stunned to learn that they were not alone and were writing to express their thanks to Raymond for showing them they weren't crazy.

After Raymond's book was reissued by a major New York publishing house, it quickly garnered a lot of attention. Over the next few years, a number of doctors, nurses, social workers, and researchers wrote to Raymond, interested in studying this phenomenon. Raymond invited them all to meet at the University of Virginia, and out of that group, four of us – psychologist Ken Ring, cardiologist Mike Sabom, sociologist John Audette, and I – formed the International Association for Near-Death Studies (IANDS) to foster research into NDEs. Talking with experiencers, seeing the aftereffects of these events on their lives, and meeting with other researchers who found NDEs equally captivating hooked me. Near-death experiences seemed the natural intersection of, on the one hand, unexplained experiences crying out for explanation; and, on the other hand, close brushes with death, which were the focus of my job in the emergency room. NDEs brought together medicine, the mind, and the thrill of scientific discovery I'd carried with me since childhood, a convergence of factors that would set the course for the rest of my career.

That journey to get to the bottom of NDEs would take me from hospital to hospital, university to university, and state to state searching for answers. Over the years, I carried out research with hospitalized patients who had come close to death through cardiac arrest, illness, accidents, suicide attempts, combat, and complications during surgery or childbirth. Almost half of these patients had lost their heartbeat, blood pressure, or breath, or had been pronounced dead. In collaboration with various colleagues over the years, I've published more than a hundred articles in peer-reviewed medical journals, describing our research findings.

In addition to research on hospitalized patients, I also gathered a sample of more than a thousand experiencers who had contacted me with accounts of their NDEs. The experiences they described were the same as those of the hospitalized patients. And so I began to collect these stories, hoping that when I had accumulated enough I'd be able to find patterns in them. And those patterns eventually led me to a better understanding of what was behind these experiences.

Outside of Time

The flames were well over two hundred feet high by the time Bill Hernlund, a twenty-three-year-old crash-rescue firefighter in the US Air Force, pulled his truck up to the rear of the burning plane. The first explosion knocked him off-balance. He fell down but was uninjured, so he scrambled back to his feet to continue fighting the fire. But then the fire ignited a second, much more powerful explosion. The burst of flames, metal, and cables launched him backward, slamming him against the side of his truck. With the second explosion, he felt pain in his head and chest, he tasted blood, and he couldn't breathe. He passed out before he hit the ground.

Bill went on to have an elaborate near-death experience. This was in 1970, years before Raymond Moody's book gave the experience a name, and when Bill had recovered and tried to share his account with his doctor, he was referred for psychiatric help. Bill then kept the NDE to himself until, almost two decades later, he discovered a local support group affiliated with the International Association for Near-Death Studies. There he learned about my interest in near-death experiences, and he wrote to me at the University of Connecticut, where I was then clinical chief of psychiatry.

Bill shared with me his account of what had happened when the burning wreckage of an airplane exploded at Ellsworth Air Force Base in Rapid City, South Dakota, and sent him flying through the air. He claimed to have seen and heard things while he was unconscious that just didn't seem possible. But I had learned not to dismiss claims out of hand just because my present worldview told me they should be impossible. And in addition to his own account, Bill also sent me newspaper clippings from the *Rapid City Journal* of April 4, 1970, with photos of the flaming aircraft and a copy of his Airman's Medal for Bravery, which cited 'his courageous action and humanitarian regard . . . with complete disregard for his personal safety.'

This is how Bill described his experience:

'I felt a lifting sensation and saw two of my buddies carrying one of the unconscious firefighters away. Somehow, I knew who the helpers were, even though they were wearing aluminized suits with hoods on, but I didn't know who they were dragging. I yelled out, "Hey, Dan, Jim, help me!" but they couldn't hear me. Then I realized that because I was the only fireman in that position, and also because my pain, taste, and smell were gone, that must be my body they were dragging away. I could see everything much more clearly and felt warm, safe, and peaceful.

'There was a roaring noise like an explosion, but duller and more prolonged. I saw Dan and Jim get knocked down on top of my body. I was in darkness, but fully conscious and vividly aware of my surroundings. I was in some kind of tunnel that looked like what a tornado funnel would be from the inside: there was a light in the distance and I saw the spiraling strings of blue-green light coming and going like the aurora borealis.

'The light was drawing me to it. I moved exceptionally fast down the tunnel and it took no time at all to reach it. It seemed that time was different or nonexistent there, wherever "there" was. The light was emanating from a being that was giving off a very brilliant light as part of his essence. He was beautiful to look at, and projected the feelings of unconditional love and peace. I also sensed other beings there, but I did not see any because I could not take my attention from the Light Being. He asked me several questions all at the same time, impressions projected at me instead of verbal word-by-word sentences. He asked: "How do you feel about your life?" and "How did you treat other people?" As he asked, every single event of my life from earliest childhood to the plane crash projected in front of me. There were details concerning people and things that I had forgotten about long ago. I was not proud of some of my dealings with other people, but the light was quick to forgive all of my errors. He told me to "be in peace" and said that my work in this world was not done yet, and that I had to go back, and I was gone.

'I was back in my body again. I do not remember traveling there. The pain was back and I smelled the odor of jet fuel and heard sirens and explosions. The doctors and medics were busy with Dan, Jim, and the B-52 crewmen, but not noticing me. Later, I found out that they looked at me long enough to see that I was dead, and turned their attention to those that they could help.

'Two days later, the doctor told me that I was lucky that I didn't die. I just said that I *did* die. He looked at me in a strange way and scheduled me for psychological evaluations. I learned to keep my mouth shut about the incident from that time on.'

Bill's experience was just one of many I'd encountered that challenged my understanding of how the world worked. But these individual accounts, taken in isolation, were difficult to study scientifically. As my interest in near-death experiences became more widely known, by word of mouth and through articles in both the scholarly and popular media, I was receiving more and more accounts to add to my collection. I set out to examine the particularly challenging features of NDEs as they presented in this growing collection of narratives, in order to find patterns that might shed light on their nature and origin.

One of the near-death-experience features that I found most puzzling was the extreme clarity and speed of thought. This is not what I would have expected of an experience that often occurs when the brain is deprived of oxygen. I was skeptical that all these experiencers could really think as clearly and as quickly as they claimed when their brains were being starved of oxygen, so I decided to look into the full range of thought processes that experiencers were describing for me. What I found is that many of them *did* report that their thoughts became much faster, clearer, and more logical than usual.

As it turns out, this is not a new phenomenon. In 1892, Swiss geology professor Albert von St Gallen Heim published the first large collection of near-death experiences in the *Yearbook of the Swiss Alpine Club*. Heim himself had had an NDE two decades earlier, when he was twenty-two and mountain climbing in the Alps. As he fell sixty-six feet down a mountain, his body crashed repeatedly against the rocky cliffs. He wrote that he had watched people fall previously, and found watching others fall to be a terrifying experience. But when he himself was falling, it was – to his shock – a beautiful experience. He reported being astounded that he was feeling no pain at all. Heim was so affected by his experience that he started talking to other climbers who had survived potentially fatal accidents, and he quickly found thirty others with similar stories. Heim described his thoughts speeding up as he fell:

'What I felt in five to ten seconds could not be described in ten times that length of time. All my thoughts and ideas were coherent and very clear, and in no way susceptible, as are dreams, to obliteration. First of all I took in the possibilities of my fate and said to myself, "The crag point over which I will soon be thrown evidently falls off below me as a steep wall since I have not been able to see the ground at the base of it. It matters a great deal whether or not snow is still lying at the base of the cliff wall. If this is the case, the snow will have melted from the wall and formed a border around the base. If I fall on the border of snow I may come out of this with my life, but if there is no more snow down there, I am certain to fall on rubble and at this velocity death will be quite inevitable. If, when I strike, I am not dead or unconscious I must instantly seize my flask of spirits of vinegar and put some drops from it on my tongue. I do not want to let go of my alpenstock; perhaps it can still be of use to me." Hence I kept it tightly in my hand. I thought of taking off my glasses and throwing them away so that splinters from them might not injure my eyes, but I was so thrown and swung about that I could not muster the power to move my hands for this purpose. A set of thoughts and ideas then ensued concerning those left behind. I said to myself that upon landing below I ought, indifferent to whether or not I were seriously injured, to immediately call to my companions out of affection for them to say, "I'm all right!" Then my brothers and three friends could sufficiently recover from their shock so as to accomplish the fairly difficult descent to me. My next thought was that I would not be able to give my beginning university lecture that had been announced for five days later. I considered how the news of my death would arrive for my loved ones and I consoled them in my thoughts. Objective observations, thoughts, and subjective feelings were simultaneous. Then I heard a dull thud and my fall was over.'

*

It struck me that it took quite a while for Heim to recount the long and involved chain of thoughts he had in just the few seconds he was falling. Many other experiencers reported the same rapid thinking. John Whitacre had an NDE at age forty-seven while recovering from surgery for pancreatic and liver cancer. He described for me his thought processes during his NDE:

'I also had the realization I had a body, which was very much like my physical body I left. I was aware of an enhanced state of consciousness, in which my mind was extremely active and alert to what I was experiencing. I was very observant during this state, and my thoughts seemed to go almost twice the normal speed, although very clear in nature.'

Among all the experiencers I've interviewed, half described their thinking during the NDE as clearer than usual, and almost as many described it as faster than usual. Gregg Nome found himself drowning at age twenty-four when his inner tube capsized after going over a waterfall. He described for me being trapped facedown in the sand at the river bottom:

'My thoughts were moving so quickly at this point. So many things seemed to be happening simultaneously, and in an overlapping manner. Suddenly, images began to pass through my mind at extremely high speeds. I was amazed to find that my mind seemed to be understanding at the same high speeds. Then, I was even more amazed at how I could be thinking of other things like this, at the same time as understanding the images. Suddenly, everything made sense. I remember thinking, "Ahhh, so that's it. Everything is so perfectly clear and simple in so many ways; I had simply never thought about it from this point of view." '

A feature associated with rapid thinking is a sense of time slowing down. Rob had an NDE at age forty-four when the ladder he was standing on slipped, tipping him over backward. He told me how his sense of time slowed down while his thoughts speeded up:

'The actual fall was slowed way down, almost like a series of camera still pictures being taken. A sort of "click," "click" visual progression. And this slowing down dramatically increased my thinking time, which resulted in my being able to size up how I could maneuver the ladder and not end up on the flagstones from two stories up. Not only did the fall slow way down, but my thinking became very clear. I actually remember wanting to head for the shrubs which, while they might pierce my skin, would break my fall. And that's exactly what happened. I rolled, avoiding head injuries. This wonderful slowing down which allowed me to think clearly in split seconds was phenomenal.'

Rob's description of time slowing, which gave him time to plan for the best way to survive, reminded me of Albert Heim's account of falling off a mountain. There may be an interesting scientific spin-off from Heim's description of his NDE. Psychologist Joe Green has raised the question of whether Heim's account of his fall played a role in Einstein's theory of relativity. Heim wrote in his 1892 article that as he fell, 'Time became greatly expanded.' In other words, time seemed to slow down for him, allowing him to think through his situation. Heim often shared his NDE account with his students at Zurich's Polytechnic Institute, where he taught geology. One of those students was a teen-age Albert Einstein, who took at least two classes from Heim that he later described in a letter to Heim's son as 'magical.' A decade later, Einstein published a revolution-ary paper describing his theory of relativity, which proposed that *time slows down the faster you travel*. It's impossible to know for certain whether this is anything more than coin-cidence, but it started me wondering whether Heim's account of time slowing as he fell faster and faster lay quietly in the back of Einstein's mind and eventually influ-enced his idea that time is not constant, but varies according to how fast you are moving.

A more extreme version of this slowing of time is an experi-ence of *complete timelessness*, a feature in many near-death experience accounts. Joe Geraci, a thirty-six-year-old police-man who almost bled to death after surgery, described this sense in his NDE:

'I knew what it was like to experience eternity, where there was no time. It's the hardest thing to try and describe to someone. How do you describe a state of timelessness, where there's nothing progressing from one point to another, where it's *all there*, and you're totally immersed in it? It didn't matter to me if it was three minutes or five that I was gone. That question is only relevant to here.'

For Joe, time not only slowed down, but seemed to disappear entirely. Many people who have had NDEs describe a sense of timelessness. Some of them say that time still existed, but that the NDE seemed to be *outside the flow of time*. Everything in their NDE seemed to be happening at once, or they seemed to move forward and backward in time. Others say that they realized in the NDE that time *no longer existed*, that the very concept of time became meaningless.

Among all the people who shared their near-death experiences with me, three-fourths reported a change in their sense of time, and more than half said that they had a sense of timelessness in their NDEs. I noticed that this slowing or stopping of time, along with the speeding up of thought processes, were more common in NDEs that couldn't have been anticipated, as in sudden car accidents or in heart attacks in apparently healthy people. They were less likely in NDEs that might have been anticipated, as in medical crises in people who knew they had a fatal disease or in people who tried to take their own lives. When these changes in thinking and the sense of time do occur, they often appear at the beginning of NDEs, and seem to be brought on by becoming aware of the threat of death. This connection between time slowing down and the suddenness of the close brush with death is something I could have discovered only by analyzing a large sample of NDEs.

The link between unexpected brushes with death and clearer, faster thinking made sense to me. If you're trying to stay alive in a sudden crisis, it may be helpful for you to be able to slow down your perception of time and to think faster and more clearly, so you might be able to save yourself, as Albert Heim and Rob did during their falls. And we know that people who are expecting to die often review their lives in anticipation of the end. Those people may not need to experience another life review when the brush with death actually comes. For these reasons, it makes sense that the shifts in thinking would be stronger in sudden and unexpected NDEs. But even though I could find reasons why people *should* think faster and clearer and slow down their perception of time in the face of life-threatening danger, it still puzzled me that they *could* do that, when I would have expected them to be terrified and hysterical. I could understand *why* time might change in an NDE, but it didn't answer the question of *how* it happens.

In addition to their *thoughts* being faster and clearer than usual, many experiencers also report that their *senses*, like vision and hearing, were more vivid than usual. Jayne Smith had an NDE at age twenty-three during a bad reaction to anesthesia during childbirth. She described for me her senses during that experience:

'I found myself in a meadow, mind cleared, identity intact, and once more aware of having a body. And this was a beautiful green meadow with beautiful flowers, beautiful colors, lit again with this glorious, radiant light, like no light we've ever seen, but there was sky, grass, flowers that had colors that I'd never seen before. And I remember so well looking at them and thinking, "I have never seen some of these colors!" And wonder of wonders, I realized I was seeing the inner light of all the growing things, just utter glory in color. It was not reflected light, but a gentle, inner glow that shone from each and every plant. Overhead, the sky was clear and blue, the light infinitely more beautiful than any light we know.'

Jayne's extraordinary sensations were visual, but sometimes other senses are involved as well. Gregg Nome, who drowned when his inner tube capsized going over a water-fall, described for me the remarkable heightening of his senses:

'Suddenly, I could hear and see as never before. The sound of the waterfall was just so crisp and clear that it really is indescribable. Two years before this, my right ear had been injured when somebody threw a large, powerful firecracker into a bar where I was listening to a band, and it exploded right next to my head. But now, in my NDE, I could hear perfectly clearly. And my sight was even more beautiful. I felt as if I had been limited by my physical senses for all these years. Sights that were very far away from me were as clear as sights that were very close, and this was at the same time. There was no blurriness in my vision whatsoever.'

Gregg found not only his vision more vivid than usual but his damaged hearing restored and all his physical senses heightened. Two-thirds of the experiencers in my research reported extraordinarily vivid sensations in their NDEs. This most often involved exceptionally bright vision and unique colors, or exceptionally clear hearing and unique sounds. On rare occasions, experiencers may report unusual odors or tastes as well.

I wasn't sure what to make of these experiences. The extraordinary thinking and perceptive abilities in NDEs, while the brain is impaired, were difficult to understand in terms of what we know about the brain. But I was drawn to their paradoxical nature, and I wanted to try to understand them. I couldn't just dismiss them. My hope was that viewing them in a larger context might help us grasp their meaning. And that larger context included a related feature of NDEs that was equally challenging.

The Life Review

Another feature of many NDEs that struck me as particularly important was the 'life review,' in which scenes from the experiencer's past come flooding back. Gregg Nome, who was drowning at age twenty-four when his inner tube capsized after going over a waterfall described being shown a rapid review of his life, including events he had long forgotten:

'I realized that I was a passive observer in the process, and it was as if someone else was running the projector. I was looking at my life objectively for the first time ever. I saw the good as well as the bad. I realized that these images were a sort of final chapter in my life, and that when the images stopped, I would lose consciousness forever.

'The images began with living color scenes of my child-hood. I was astonished, because I saw myself sitting in a baby's high chair and picking up some food with my hand and throwing it onto the floor. And there was my mom, twenty-five years younger, telling me that good boys don't throw their food on the floor. Next, I saw myself at a lake on a summer vacation we took when I was about three or four years old. My older brother and I had to swim with an air bubble on our backs to help us float, because neither of us was able to swim on our own yet. For some reason I was mad at him, and to demonstrate my point, I threw his air bubble into the lake. Mark was very upset and began to cry, and my father walked over and explained to me that it wasn't nice to do what I did, and that he and I would have to row out in the boat to get it, and I would have to apolo-gize. I was amazed at how many scenes I was seeing that had long since been forgotten.

'It seemed that all of the scenes had to do with experi-ences that I had learned from in some way as I matured. I also saw events that were traumatic in various ways. The images continued at high speeds, and I knew that time was about to run out, for the images were getting closer and closer to the present. Then the images ceased. There was only darkness, and a feeling of a short pause, like some-thing was about to happen.'

As I looked for research that had already been done on life reviews, I discovered that this was also not a new phe-nomenon. In 1791, when British rear admiral Sir Francis

Beaufort was only a seventeen-year-old midshipman, he fell off a boat into Portsmouth Harbor on the southern coast of England. Unfortunately, he had not yet learned to swim. After exhausting himself struggling to breathe, he lost consciousness and immediately experienced a feeling of calmness and noticed changes in his thinking. He later described it this way:

'From the moment that all exertion ceased – which I imagine was the immediate consequence of complete suffocation – a calm feeling of the most perfect tranquility superseded the previous tumultuous sensations – it might be called apathy, certainly not resignation. Though the senses were thus deadened, not so the mind; its activity seemed to be invigorated, in a ratio which defies all description, for thought rose above thought with a rapidity of succession that is not only indescribable, but probably inconceivable by anyone who has not himself been in a similar situation.

'The course of those thoughts I can even now in great measure retrace – the event which had just taken place – the awkwardness that had produced it – were the first series of reflections that occurred. They took then a wider range – our last cruise – a former voyage, and shipwreck – my school – the progress I had made there, and the time I had mis-spent – and even all my boyish pursuits and adventures. Thus traveling backward, every past incident of my life seemed to glance across my recollection in retrograde succession; not, however, in mere outline, as here stated, but the picture filled up with every minute and collateral feature. In short, the whole period of my existence seemed to be placed before me in a kind of panoramic review, and each act of it seemed to be accompanied by a consciousness of right or wrong, or by some reflection on its cause or its consequences; indeed, many trifling events which had been long forgotten then crowded into my imagination, and with the character of recent familiarity.'

Beaufort described his thoughts not only speeding up but encompassing every single incident in his life and judging every action as right or wrong. Many of the experiencers who shared their stories with me described this kind of life review.

Tom Sawyer, a thirty-three-year-old supervisor in the town highway department, had an NDE when a truck he was working under came crashing down on his chest. I first met him through a letter he sent me in 1981. He wrote that he had promised Raymond Moody's wife, Louise, that he would contact me to volunteer to take part in my research. I would get to know Tom and his wife, Elaine, quite well over the next twenty-five years, until he finally succumbed to chronic lung disease. But over that quarter century, I never lived more than a short day's drive from his home, and he found frequent reasons to make the trip. He described for me in vivid detail the accident that caused his NDE:

'I was working on my pickup truck. My nine-year-old son, Todd, was home from school and wanted to help. I was being very careful to practice safety procedures, especially for Todd's benefit. I had the truck jacked up safely with blocks and timbers and jack stands. I had it jacked up and I was lying on my back on a mechanic's creeper. I was going to replace a tie-rod end and repair or rebuild the transmission linkage.

'I rolled underneath and told Todd what tools I needed to get started with my work. I began working on the transmission linkage and suddenly the truck started to move! At the first instant it moved, I knew something was terribly wrong. I had done more than what is expected to make the truck safe before crawling underneath it. I found out more than a week after the incident that the driveway underneath the jack had given way. There was an air pocket underneath the asphalt, and it gave way under the jack, which moved the truck sideways. The sideways motion tipped over the six-by-six timbers, and as the truck came down, it missed the jack stands in front of the truck and of course plunged into me.

'The truck dropping down onto my body seemed to be in extreme slow motion. As it fell, I made an attempt to yell out, "Todd, get help." But the four-thousand-pound truck came down on my chest before I got much out; it squeezed all the air out. The frame of the truck came across the center of my chest between the bottom rib and my breast.

'Having the air suddenly squished out with only half a breath in me, the probability is that I would not have been able to hold my breath for long. I shook my head and tried to fight off the unconscious state with a fervent desire to survive this stupid accident and get on with things. Ultimately I did run out of oxygen, and kind of drifted off. I realized I'd become paralyzed and then eventually about the last thing that happened was that my eyelids felt like they were closing and I lost eyesight. At that time my heart was still beating its last several beats, getting slower and slower on the last few beats. It was very curious and intriguing to experience the last three heartbeats. Then there was just a kind of a blankness.'

Tom described, at some point during the NDE that followed, reliving painful incidents from earlier in his life:

'I experienced a total life review. The best way to describe it is to give you an example. When I was around eight years old, my father told me to mow the lawn and cut the weeds in the yard. Aunt Gay, my mother's sister, lived in the cottage out back. She was always fun to be with. Certainly all the kids thought she was a cool person to know. She had described to me her plans for some wildflowers that grew on little vines in the backyard. "Leave them alone now, Tommy," she said.

'However, my father told me to mow the lawn and cut the weeds. Now, I could have explained to my father that Aunt Gay wanted the weeds left to grow in this particular area. Or I could have explained to Aunt Gay that Father had just told me to mow the lawn and said to cut that patch of weeds. Or, I could methodically and deliberately go ahead and mow the yard and cut the weeds. I did that. I deliberately decided to be bad, to be malicious. My Aunt Gay never said a word to me; nothing was ever mentioned. I thought, "Wow, I got away with it." End of story.

'Guess what? I not only relived it in my life review, but

I relived every exact thought and attitude; even the air temperature and things that I couldn't have possibly measured when I was eight years old. For example, at the time, I wasn't aware of how many mosquitoes were in the area. In the life review, I could have counted the mosquitoes. Everything was more accurate than could possibly be perceived in the reality of the original event. I experienced things that cannot be perceived. I watched me mowing the lawn from straight above, anywhere from several hundred to a couple of thousand feet, as though I were a camera. I watched all of that. My life review was absolutely, positively, everything basically from the first breath of life right through the accident. It was that panoramic view. It was everything.'

I had heard other experiencers describe reliving their lives in exquisite detail, and I could understand that as a psychological reaction to the threat of death. But then Tom went on to describe an additional feature of his life review that was harder for me to understand. Tom relived his entire life not only through his *own* eyes, but also from the perspective of *other people*. He described this aspect graphically:

'I not only reexperienced my eight-year-old attitude and the kind of excitement and joy of getting away with something, but I was also observing this entire event as a thirty-three-year-old adult, with the wisdom and philosophy I was able to attain by that time. *But it was more than that.*

'I also experienced it exactly as though I was my Aunt Gay when she walked out the back door and saw the weeds had been cut. I knew the series of thoughts that bounced back and forth in her mind. "Oh my goodness, what has happened? Oh well, Tommy must have forgotten what I said. But he couldn't have forgotten— Oh no, knock it off. Tommy's never done anything like that. Gee, it was so important. He had to know . . . He couldn't have known."

'Back and forth, back and forth, between thinking of the possibility, and saying to herself, "Well, it is possible. No, Tommy isn't like that. It doesn't matter anyway; I love him. I'll never mention it. God forbid, if he did forget and I remind him, that will hurt his feelings. Should I confront him with it and just ask him?"

'What I'm telling you is, I was in my Aunt Gay's body, I was in her eyes, I was in her emotions, I was in her unanswered questions. I experienced the disappointment, the humiliation. It was very devastating to me. It changed my attitude quite a bit as I experienced it.

'In addition to this, and what is probably more important, spiritually speaking, I was able to observe the scene, absolutely, positively, unconditionally. In other words, not with the horrendous emotional ill-feelings that my Aunt Gay experienced. I experienced it with this unconditional love that is only God's eyes, or the eyes of Jesus Christ, or the light of Jesus, or the light of Buddha enlightened, the spiritual entity. No judgmental aspect whatever. This is simultaneous with the total devastation of what I created in my aunt's life. And the arrogance, the snide little thoughts, the bad feelings, and the excitement of what I created in my own life at that young age.'

*

Among all the participants in my research, a quarter reported a life review. Some experiencers told me that their entire lives flashed before their eyes, from birth to the present or in reverse order. Others said that they were able to view different scenes from their lives at will. The vast majority described this life review as more vivid than ordinary memories. Some experiencers told me that they were shown images from their past, as on a movie screen or on pages in a book. But many, like Tom, reported that they *reexperienced these past events as if they were still happening*, with all the original sensations and feelings.

Three-fourths of those who had a life review said that it changed their ideas of what things are important in life. Half of those who had a life review experienced a sense of judgment, most often judging themselves, about the right-ness or wrongness of their actions. And more than half experienced these past events not only through their own eyes, but also – like Tom – from the viewpoints of others, feeling those other people's emotions as well as their own.

Barbara Harris Whitfield had an NDE at age thirty-two when she suffered respiratory complications while immobilized after back surgery. She described a life review in which she reexperienced abusive childhood events from the perspective of other people involved:

'As I left my body, I again went out into the darkness. Looking down and off to the right, I saw myself in a bubble – in the circle bed – crying. Then I looked up and to the left, and I saw my one-year-old self in another bubble – facedown in my crib – crying just as hard. I decided I didn't want to be the thirty-two-year-old Barbara anymore; I'd go to the baby. As I moved away from my thirty-two-year-old body in the circle bed, I felt as though I released myself from this lifetime. As I did, I became aware of an Energy that was wrapping itself around me and going through me, permeating me, holding up every molecule of my being.

'In every scene of my life review I could feel again what I had felt at various times in my life. And I could feel everything everyone else felt as a consequence of my actions. Some of it felt good and some of it felt awful. All of this translated into knowledge, and I learned – oh, how I learned! The information was flowing at an incredible breakneck speed that probably would have burned me up if it weren't for the extraordinary Energy holding me. The information came in, and then love neutralized my judgments against myself. I received all information about every scene – my perceptions and feelings – and anyone else's perceptions and feelings who were in the scene. There was no good and no bad. There was only me and my loved ones from this life trying to be, or just trying to survive.

'I went to the baby I was seeing to my upper left in the darkness. Picture the baby being in a bubble and that bubble in the center of a cloud of thousands and thousands of bubbles. In each bubble was another scene in my life. As I moved toward the baby, it was as though I was bobbing through the bubbles. At the same time there was a linear sequence in which I relived thirty-two years of my life. I could hear myself saying, "No wonder, no wonder." I now believe my "no wonders" meant "No wonder you are the way you are now. Look what was done to you when you were a little girl."

'My mother had been dependent on drugs, angry, and abusive. I saw all this childhood trauma again, in my life review, but I didn't see it in little bits and pieces, the way I had remembered it as an adult. I saw and experienced it just as I had lived it at the time it first happened. Not only was I me, I was also my mother. And my dad. And my brother. We were all one. I now felt my mother's pain and neglect from her childhood. She wasn't trying to be mean. She didn't know how to be loving or kind. She didn't know how to love. She didn't understand what life is really all about. And she was still angry from her own childhood, angry because they were poor and because her father had grand mal seizures almost every day until he died when she was eleven. And then she was angry because he left her.

'Everything came flooding back. I witnessed my brother's rage at my mother's abuse, and then his turning around and giving it to me. I saw how we were all connected in this dance that started with my mother. I saw how her physical body expressed her emotional pain. I could hear myself saying, "No wonder, no wonder." I could now feel that she abused me because she hated herself.

'I saw how I had given up myself in order to survive. I forgot that I was a child. I became my mother's mother. I suddenly knew that my mother had had the same thing happen to her in her childhood. She took care of her father during his seizures, and as a child she gave herself up to take care of him. As children, she and I both became anything and everything others needed. As my life review continued,

I also saw my mother's soul, how painful her life was, how lost she was. In my life review I saw she was a good person caught in helplessness. I saw her beauty, her humanity, and her needs that had gone unattended to in her own childhood. I loved her and understood her. We may have been trapped, but we were still souls connected in our dance of life by an Energy source that had created us.

'As my life review continued, I got married and had my own children and saw that I was on the edge of repeating the cycle of abuse and trauma that I had experienced as a child. I was becoming like my mother. As my life unfolded before my eyes, I witnessed how severely I had treated myself because that was the behavior shown and taught to me as a child. I realized that the only big mistake I had made in my life of thirty-two years was that I had never learned to love myself.'

How do we make sense of a life review? For the past half century, 'life review therapy' – a guided, systematic, thorough review of major life events – has been a major tool for counselors working with people at the end of their lives. It can help people cope with loss, guilt, conflict, or defeat, and find meaning in their lives and in their accomplishments. This closure can be critical in helping people face death more peacefully.

And for people who have NDEs and then return, their life review can not only help them cope with losses and find meaning in their lives but also help them make changes in their behavior based on what they've learned. Tom's reexperiencing his life events, not only through his eyes but also from the perspective of others, helped him understand the pain he caused others and avoid repeating that behavior. Barbara's reliving her childhood trauma, not only through her experience but also from her mother's life, helped her understand and come to terms with her own abuse, and to make changes in her own life to avoid perpetuating that abusive cycle with her own children.

In addition to all these recent reports of life reviews, we also have to keep in mind the similar examples from past centuries, such as Rear Admiral Beaufort's account of his NDE in the late eighteenth century. Like Albert Heim's description in the late nineteenth century of his thoughts speeding up and time slowing down as he fell off a mountain, these historical narratives suggest that NDE accounts are not just reflections of our current model of what happens when we die. They have challenged our ideas of how the mind and brain work for hundreds of years.

Knowing that they are not new phenomena, but perhaps universal experiences that people have been having for centuries, doesn't tell us what they are. Are they due to common psychological mechanisms that help us gain some closure before we die? Are they caused by brain malfunctions as we start to approach death? Or are they something else entirely? At that point, I didn't have the tools I would need to study NDEs more thoroughly. So I set about trying to develop a more systematic way of organizing and analyzing them than just collecting these stories. And this would introduce a whole new set of questions and challenges.

4

....

Getting the Whole Story

Word that I was interested in near-death experiences continued to spread, and I started hearing from more and more people who offered to share their own accounts with me. I knew that the more NDEs I could collect, the easier it would be to identify the images and features of the experience that kept cropping up repeatedly. And the more information I could gather about the specific medical details of their brushes with death, the easier it might be to identify the biology associated with these experiences. But I also realized that the NDEs these people were offering me could be a biased sample of all the NDEs out there. These were self-selected – the accounts of experiencers who were both able and willing to share their stories. Were they different from other NDEs – the ones that people didn't offer to share, or simply couldn't put into words?

I decided that, in addition to gathering accounts from those self-selected 'volunteers,' I would also have to interview a large number of people who had come close to death, but who hadn't brought their experiences to me. With my position at the university hospital, I had access to just such a group. With the approval of the cardiology department, I set up a study to interview everyone admitted to my hospital because of a serious heart problem. In a period of two and a half years, I interviewed almost 1,600 patients who were admitted to the inpatient cardiac service, of whom 116 had had a cardiac arrest, in which their hearts stopped completely, documented in their medical records.

Claude, a seventy-two-year-old farmer, was one of those whose heart had stopped. The day after he was admitted, I made my way to his hospital room, introduced myself, and asked if he'd be willing to talk with me about what happened to him. He gave me a puzzled look, as if it was perfectly obvious what had happened to him. But he agreed to talk. I told him I understood that his heart had stopped, and I asked him, as I ask every patient, 'What was the last thing you remember before you blacked out?'

'I was slopping the hogs,' Claude began slowly, 'and I started feeling dizzy, so I walked back to the barn and sat down on a bale of hay.' He paused, and then added, 'And that's the last thing I knew.'

'And what was the next thing you remember after that?' I asked.

'I woke up in this bed, with wires on my chest and a tube in my arm, and I don't know how the heck I got here.'

Trying to sound matter-of-fact, I asked a third question

that I put to all these patients: 'And what do you remember in between those two times?'

Claude hesitated, as if he was sizing me up, and then said, just as matter-of-factly, 'I thought I was going to meet my maker, but my paw – he's been gone maybe fifteen years now – stopped me and said I had to go back.'

I kept my voice calm and professional, although my heart started racing with excitement at hearing an NDE account from an unbiased source. I leaned in and, nodding encouragingly, said, 'Tell me about meeting your paw.'

Claude looked at me patiently, and after a very brief pause, he said, 'I just did.'

I nodded and tried to figure out how to word a follow-up question. But Claude closed his eyes and said, 'I'm tired. That's all I have to say.'

That was enough to tell me that there would be others in this hospital population like Claude who had NDEs. And in fact, I found twenty-six other cardiac patients who shared with me their NDE accounts. It turned out that 10 percent of the patients whose hearts had stopped described NDEs, as well as 1 percent of the other patients, who had had heart attacks or other serious cardiac events without complete heart stoppage.

Now I was faced with how to evaluate these accounts of NDEs. Of course, I couldn't observe their experiences directly. All I knew was what the experiencers told me and how they were affected by their NDEs. But one of the first things many experiencers say is that there are no words to describe what they experienced. So when I follow that by asking them to tell me about it, I find that I'm asking them to do something very difficult. Many experiencers turn to whatever cultural or religious metaphors they have available in order to describe things that don't have familiar labels. For example, many Americans who have NDEs describe passing through a long, dark space they call a 'tunnel.' Experiencers from less-developed countries where tunnels are rare may label it a 'well' or a 'cave.' Dominic, a truck driver who had an NDE when his eighteen-wheeler collided with another vehicle on the interstate, described traveling through a long, dark 'tailpipe,' which was the image that came most readily to him.

Many experiencers feel frustrated by this difficulty putting the experience into words. Joe Geraci, the policeman who almost bled to death after surgery, described for me his frustration at trying to share his NDE:

'It's impossible to describe. It truly can't be put into words. It's the hardest thing to try and describe to someone. It certainly didn't fit any stereotypes. I guess what I'm trying to say is, there is nothing I can draw reference to in my life to begin to explain it. It's frustrating to try and talk about it. I'm trying to verbalize to you something I can't even verbalize to myself. It's so simple and so deep, that that's the problem.

'It's frustrating. It really can't be expressed the way it is; I always fall short. I'm falling short right now. So whatever I say to someone goes into their little sifter, through their own experiences, and registers based on their own frame of reference. I wanted to tell my wife, but I literally couldn't speak. It's hard to experience a thing so beautiful, something that means so much to you, something that's changed your life, and be so darn lonely at the same time.'

Likewise, Bill Urfer, a forty-six-year-old businessman, related to me his difficulty describing the NDE he had during abdominal surgery for a ruptured appendix:

'What I say here is limited by the English language, for no words have been invented to tell this story with adequate beauty. Nothing in my life could have prepared me for what I had seen. I know the futility of trying to describe the scene, yet I keep trying. No words came that could begin to describe what I had seen. I had to share it with somebody, but I couldn't find words to describe what I had seen. Again and again, the thoughts pounded through my head, at times seeming ready to spill out for everyone to see. Searches through a dictionary were futile, the words bland, lacking the fire of color.

'Try to draw an odor using crayons. You can't even begin to try, no matter how many crayons you have in your box. That's what it's like describing NDEs with words. No matter how many words you use, you can't really describe what an NDE is like. Lying awake in the dark, I tried forming sounds that would explain. Maybe music would do what speech could not. After all, no one can describe the beauty of certain sounds, sounds that move us to action or to tears. Yes, maybe music was the only form of communication that could explain the feelings of peace that never left.'

*

And Steve Luiting, who had an NDE when he drowned at age eight, described the difficulty of trying to describe his experience like this:

'The language spoken after death was much, much more complex and could literally encapsulate experiences. Even the memories when coming back into my body flattened, simplified, and became symbols of what really happened. I believe this flattening happens simply because the human brain can't understand a world so much more complex and possibly so alien. When I read about people having seen streets of gold it is amusing, because that would be a flattened example of a complex visual reference – not gold so much as vibrant and alive, I guess.'

Steve's inability to find adequate ways to describe his experience led him to regard other experiencers' concrete descriptions not as literal but as metaphoric. He saw descriptions of streets of gold, pearly gates, and angelic figures as the best analogies others could come up with to convey what is essentially an indescribable experience. Jalāl ad-Dīn Rumi, the thirteenth-century Sufi mystic, wrote that 'Silence is the language of God; all else is poor translation.' That's what many experiencers seem to be telling us about their NDEs. And most experiencers are less skillful with words than Joe, Bill, and Steve. Many of them, like Claude, are unable – or perhaps unwilling – to describe their NDEs in any detail. That said, I needed some systematic way to talk about NDEs in order to carry out scientific research and to come to some logical understanding of them.

Communication is always difficult, as words often seem inadequate, and that is particularly true in communicating intense emotional experiences. But the difficulty many experiencers encounter when searching for words isn't the only thing that holds them back from sharing their stories. Some people who've had near-death experiences are afraid – and rightfully so – that they will be thought of as crazy, or as lying. An example of that fear was Gina, a police officer who had tried to kill herself. I met her in a study I was carrying out interviewing patients who were hospital-ized after attempting suicide. My hope was to discover what experiences they might have had during their brushes with death, and then to follow up with them monthly to reassess their continuing thoughts about suicide. I wanted to know whether having a near-death experience would change people's attitudes toward taking their lives.

Gina was a twenty-four-year-old rookie officer. Only five feet, two inches tall with a petite build and somewhat unruly black curls, she nevertheless had a toughness and grit that let people know she meant business. She had loved her classes and training at the police academy, but once she joined the force she was embedded in a macho culture where she felt ridiculed and harassed. Being a police officer was all she had wanted for years, and when her sergeant began aggressively teasing and touching her, she felt she was in an impossible situation. Feeling trapped and unable to see a way out, she took a serious overdose and found herself in the psychiatric ward. I guessed that her overdose was at least in part a cry for help, whether conscious or not. If her goal had been only to kill herself, using her service revolver would have been a more certain method, and more in keeping with her no-nonsense character.

I asked Gina my usual questions: What was the last thing she remembered before losing consciousness? What was the next thing she remembered after that? What did she remember between those two times? She denied having had any experiences while she was unconscious. I therefore considered her a participant in my 'control group' – the suicide attempters who *didn't* have NDEs. But when I contacted her a month later to see how she was doing and to reassess her suicidal thoughts, she surprised me.

'Gina,' I began, 'do you remember talking to me after you woke up from your overdose?'

'Yeah,' she replied, with a bit of hesitation. 'You asked what I remembered about it, and I didn't tell you.'

I raised an eyebrow. 'You didn't tell me . . .?'

She paused again, then went on. 'While the paramedics were transporting me to the hospital, I left my body.'

Now I paused, wondering how to respond to this. Was she making something up to please me, because she thought I'd been disappointed in her response a month earlier? I decided to give her the benefit of the doubt, at least for the time being. 'You didn't *remember* that last month, or you weren't comfortable sharing it with me at that time?'

She nodded, her forehead still wrinkled in concern. 'No, I wasn't sure you were taking this seriously, so I didn't say anything then.'

'Well, what can you tell me about it now?'

Now she launched into her story readily. 'I was against the side of the ambulance, looking at my body and the paramedic sitting next to me. He was adjusting the flow of liquid into the IV in my arm. He seemed bored, and not very concerned about me. But then, I wasn't concerned, either. As I watched him do that, and saw my body not moving at all, I thought, "Well, that's interesting," and that was all. I didn't feel any particular attachment to my body, any more than I did to *his*.'

I waited for her to go on, and then asked, 'What else?'

She paused, and then shook her head. 'That's it.'

I finished the interview by asking her the standard questions about her suicidal thinking, and how her life was going. She told me she had confronted her sergeant about his harassment. He acted as if he didn't know what she was talking about, so she complained to the police chief about the sergeant. She remained in the same position, but the harassment seemed to have stopped. I told her I thought that had been the right thing to do, though it must have taken guts. I asked whether she had any questions – she didn't – and thanked her for speaking with me.

I met with her a third time a month later. 'Gina, when we spoke last time you told me about leaving your body in the ambulance after your overdose.'

'Yeah,' she said with an embarrassed smile, 'but I didn't tell you about seeing my cousin.'

My eyebrow went up again. 'Your cousin?'

'Yeah,' she continued, not looking at me, 'my cousin Maria was there with me in the ambulance. She died four years ago in a car accident. We were the same age, and we did everything together. She told me that I still had lots I needed to do, and that I had other choices besides ending my life. She was a bit sarcastic, like she always was, but she was also sad that I'd overdosed.'

She paused, then went on. 'She told me she was sending me back so I could confront my sergeant and not let him get away with that. And she said that if I tried to kill myself again, she would only kick my sorry butt back here again.'

'You didn't feel safe telling me about that last month?'

She looked me in the eye now and laughed. 'You're a shrink, for God's sake! I didn't want to wind up back in the hospital! I wasn't going to risk you thinking I was nuts!'

I nodded and laughed along with her. 'But you're okay with telling me about this now?'

She turned serious, still looking at me. 'Well, you didn't have me committed for saying that I'd left my body, so I figured I could probably trust you with this, too.'

We talked for a while about her memories of her cousin and what it had been like to have her cousin send her back, and she finally seemed to run out of steam. Once more, I ended with the standard questions about suicide and how she was doing. She gave a deep sigh and said that she didn't think the police chief had taken her seriously. She had gone on to talk with her union representative and she'd filed a formal complaint, and then wrote a letter to the city district attorney. I again reinforced her decision to take action, gave her a chance to ask me any questions, and thanked her for speaking with me again.

When I tried to contact her after another month had passed, I learned that she had quit her job with the police force and had told her supervisor that she was moving back to her hometown. I tried to track her down to interview her again, but I wasn't able to find her.

Of course, it's possible that she didn't tell me about leaving her body and meeting her cousin when we first spoke after her overdose because it didn't really happen. It's possible that she was making things up each time we met. But she had no obvious reason to invent material; her emotional reactions seemed genuine. Whether she remembered her experience accurately is another matter, and not something I could verify. But it rang true that Gina would have been reluctant to tell a psychiatrist about her experience – especially on our first meeting, when she was trying to get discharged from the hospital. Was there even more to her NDE that she still hadn't felt comfortable sharing? I'll never know.

I found in interviewing people after a near-death experience that there are a number of reasons for experiencers to keep their NDEs to themselves. Remember, these are often earth-shattering experiences. Some people who've had NDEs are so shaken by the experience that they're not ready to talk about it. Some are depressed or angry at having found themselves back in their bodies. Some are confused about conflicts between what they experienced and what their religion taught them to expect at death. Some fear that their NDE is a sign that they are mentally ill – or that other people will take it as evidence. And some experiencers whose NDEs happened during an assault, a suicide attempt, or an avoidable accident don't want to talk about the experience because they're too traumatized by the event, or feel ashamed or blame themselves.

Many experiencers worry that other people, including researchers, won't understand what happened to them. They fear they'll be ridiculed if they talk about their NDE. Some feel that sharing it with other people will sully or trivialize it. And some feel their near-death experience is too personal to share. They believe that the information they received in their NDEs was given to them for their own benefit, and is not meant to be studied or analyzed by the scientific community.

It can be difficult for researchers – as well as family and friends – to know if an experiencer is giving them the whole story. There are so many reasons for patients *not* to be willing to share their experiences that I am always grateful when they *do*. After an NDE, experiencers can be very vulnerable, and what they do *after* their experiences can be important to their future well-being.

5

....

How Do We Know What's Real?

In 1978, not long after I'd met Raymond Moody and started exploring near-death experiences with Ian Stevenson, I realized that I would need more training in medical research skills than I could get at a clinically oriented med school like the University of Virginia was at that time. I took a position on the faculty of the University of Michigan, a major research medical school, where there were senior mentors who could teach me what I needed to know in order to study NDEs with scientific rigor. I was particularly lucky to have the late Gardner Quarton, director of the Mental Health Research Institute at Michigan, take me under his wing and teach me how to develop practical research questions and design solid research protocols.

Much of the early NDE research, including mine, consisted of little more than collecting first-person accounts, with no consistent format from one researcher to another. Several of my colleagues at other institutions were gathering the information they thought was critical to the experience. Researchers who were interested in clarity of thought during NDEs focused on time distortion, life reviews, and so on, but didn't ask about the sensation of leaving the body or meeting deceased loved ones. Others who were interested in the religious implications focused on visions of God and an afterlife, but didn't ask about changes in mood or thinking. As I reviewed these accounts, it was difficult to know whether we were all collecting examples of the same kind of experience, or whether we were studying a variety of different experiences that might happen when people think they're dying.

One of my colleagues defined near-death experiences as 'anything people experience when they're close to death.' But that seemed to me too broad. There are many kinds of experience people have when faced with death, from blacking out to panicking to acceptance – which are all very different and not at all what Moody meant by the term 'NDE.' I realized we needed a way to put us all on the same page when talking about near-death experiences. This was a challenge. In addition to the personal biases of different researchers, each of us was acting in relative isolation, unaware of who else might be studying NDEs or how others were defining the experience. I wanted to bring some logical order to the study of this experience.

To tackle this problem, I developed the NDE Scale in the early 1980s as a way to standardize what we mean by the term 'near-death experience.' I started with a list of the eighty features most often mentioned in the literature on NDEs, and sent this list to a large sample of experiencers. Then, through a series of repeated assessments by experiencers and other researchers, with the help of statistical analyses, I whittled the scale down to a more manageable list of sixteen features. These sixteen features included changes in thinking, such as thoughts speeding up and the reviewing of scenes from the past. They included changes in emotions, such as a feeling of intense peace and a feeling of unconditional love often radiating from a being of light. They included extraordinary perceptions, such as an awareness of things going on elsewhere and a feeling of separating from the physical body. And they included a sense of 'otherworldly' experiences, such as seeing deceased loved ones or religious figures and coming to a border or 'point of no return.'

Experiencers could rate each of these sixteen features 0, 1, or 2 points, giving a total score between 0 and 32. For example, Bill Hernlund's experience when he was caught in the airplane explosion received a score of 28, and Tom Sawyer's NDE when his truck crushed his chest scored 31. These scores are helpful in comparing research across different investigators, but they are not useful in dealing with individual experiences. I've found that many experiences with low scores on the scale nevertheless lead to life-changing spiritual transformations. So the NDE Scale is not a measure of how deeply an experiencer may be affected. It's simply a tool that researchers can use to make sure they're investigating the same experience. And in the thirty-eight years since it was first published, the NDE

Scale has stood the test of time, having been translated into more than twenty languages and used in hundreds of studies around the world.

I was a bit surprised to discover, after I went through the rigorous procedure to create the scale, that it turned out not to include some things that are common in NDEs, like going through a tunnel. People *do* report going through a tunnel in NDEs, but they also report going through a tunnel in many other kinds of experience. Some researchers have suggested that the sense of going through a tunnel is something our minds imagine, in order to explain to ourselves how we got from one situation to another, when we aren't consciously aware of how we made that transition. This has been compared to what theoretical physicists call a 'wormhole' that connects one universe to another. I'm not sure that's the best explanation for the tunnel, but the fact remains that people experience tunnels along with the other NDE features about as often as they experience tunnels *without* those other features. So going through a tunnel is not something that researchers can use to distinguish NDEs from other experiences people may have when they come close to death.

Twenty years after this scale was published, and long after it had become accepted as the standard tool of NDE researchers worldwide, I was challenged by two skeptical scholars I didn't know: Rense Lange, a statistician from Southern Illinois University School of Medicine, and Jim Houran, a psychologist then at the University of Adelaide in Australia. These scholars had no previous interest in NDEs, but were applying a complicated statistical test to various scales that had been developed by other researchers – and in the process 'debunking' some of them. They wanted me to give them the raw responses on the scale that I had collected from around three hundred people who had come close to death, and let them carry out their sophisticated statistical test on the data to see whether the NDE Scale was valid.

Apprehensive, I had reservations about working with them. I'd already put years of work into this scale, and it had become accepted by scholars around the world. I wasn't familiar with the statistical test they wanted to carry out. I didn't know whether it was a good test, and whether my scale would hold up under it. What if the scale failed the test? Would it cast doubt on all my work with NDEs? Would it ruin my credibility and my career as a scientist?

On the other hand, if the NDE Scale was faulty, I certainly wanted to know that! How could I refuse to share my data and put my scale to the test? If I was truly a skeptic, how could I be skeptical of other people's ideas but not my own? I'd met too many academics who called themselves 'skeptics' but refused to look at any evidence that might challenge their own beliefs. Could I swallow my pride – and my fear of failing – and expose my data to an independent test? That's what intellectual honesty required. That's what a true skeptic would do. That's what my father, had he still been alive, would have wanted me to do. I handed over all my data on the NDE Scale, the responses of hundreds of people who'd had near-death experiences, and waited for the results from Rense and Jim. As the months went by, I had many fitful nights second-guessing my decision to subject my work to that scrutiny. But each morning, in the light of day, I knew that it was the right thing to do.

To my great relief, their analysis ended up confirming the validity of the NDE Scale. It showed that the scale measured one consistent experience that was the same for men and women and for people of all ages, across many cultures. NDE Scale scores were the same no matter how many years had passed since the experience. I heaved a huge sigh of relief. My NDE Scale – and by extension NDEs themselves – had been given the stamp of credibility by a team of skeptics who not only had no stake in near-death experiences, but would have been happy to discredit them.

While I was at the University of Michigan as chief of the psychiatric emergency service, in the evenings and week-ends when not with my family I continued working with Ian Stevenson back in Virginia long-distance by telephone and letter – this was long before the days when we had per-sonal computers, let alone email.

In 1979, our work reached a pivotal moment. Ian and I published a short article on near-death experiences in the *Journal of the American Medical Association* (*JAMA*). We made the point in that article that, despite the increasing number of books and articles about death and dying in recent decades, their authors consistently ignored the ques-tion of whether or not our consciousness might continue after death. We did *not* claim that NDEs provided evidence of consciousness after death, but rather that they may shed some light on our ideas of some kind of continued consciousness. We noted that experiencers' expectations before they approached death might influence how they make sense of the type of NDE they report, but also that NDEs often contradict experiencers' beliefs about an afterlife. Furthermore, we described features that were consistent across different countries and societies. Some of those features run counter to many cultural or religious beliefs. Finally, we noted that experiencers almost universally become convinced by their NDEs that some part of them will live on after death. We concluded that there were many aspects of NDEs that had not yet been explained and deserved further research.

Up to that point, few of my colleagues knew of my near-death research. My work hours were taken up mostly with treating patients and teaching medical students. I still had mixed feelings about NDEs. On the one hand, they smacked of religion and folklore, things that were foreign to my scientific, 'what you see is what you get' upbringing. I couldn't make sense of NDEs in terms of physical particles and forces, so how could they be real?

On the other hand, they happened. And many people not only reported them but considered them positive, life-changing experiences. The article that Ian and I published brought me out of the academic closet, and I was as surprised as I was gratified that it had been accepted for publication. In fact, I was thrilled that I now had a paper published in the second most widely read medical journal in the world. Even though my colleagues would now know about my unusual interest, it had the thumbs-up from one of the medical establishment's most respected journals.

My elation didn't last long. A couple of months after our article was published, I received a letter from Ian. Enclosed with his message was a letter to the editor that had been submitted to *JAMA*, complaining about the publication of our paper on NDEs. The letter writer, the chairman of orthopedic surgery at a hospital in New York, argued that NDEs were the concern of religion and not doctors, and had no business being mentioned in a medical journal. *JAMA*'s editor, who had been steering the journal toward providing practical information for practicing doctors, had forwarded the letter to Ian, inviting us to write a response for publication along with the complaint.

I was intimidated by this letter. I felt I was in way over my head, as if I was being slapped down for daring to think I could play with the big boys. Part of me wanted to fight back, but another part felt like apologizing for my audacity and slinking away into the night. I was terrified that, when this letter to the editor from a department chairman eventually appeared in print, my career and reputation would be ruined. Fortunately, Ian, a former department chairman himself, was not intimidated in the least. Instead, he took the lead in crafting a response for us, in which we pointed out how important it was for doctors to know about NDEs and to take them seriously.

For one thing, NDEs often happen to people under medical care for serious illnesses and injuries. At the time, we knew very little about the physiological changes taking place under such circumstances that might be associated with NDEs, and we could learn about them only if doctors were more aware of and interested in studying NDEs. For another thing, NDEs usually change the experiencers' beliefs about death and dying, which can have profound effects on their lifestyles and attitudes toward medical treatment. Surely, we argued, doctors who care for their patients would want to understand these experiences and their impact.

The complaining letter and our response were published together in *JAMA* six months after our original article. That seemed to be the end of the matter. There were no further letters to the editor, and none of my colleagues at Michigan ever acknowledged either the original article or the subsequent exchange. I ended up feeling emboldened by the whole incident, and in the next few years I went on to publish several additional articles in major psychiatry journals describing my near-death research.

I also felt comfortable enough a couple of years later to organize a panel on near-death experiences at the annual meeting of the American Psychiatric Association. But the night before I was to give my presentation on NDEs, I had a terrifying dream. I felt my body getting larger and larger. At first, there was no particular emotion associated with that sensation, but I kept growing, and soon became bigger than the planet Earth. As I continued to expand through the universe, reaching toward the distant stars, I suddenly realized that the atoms that made up my body had not grown in size – I kept getting larger because the distance between my individual atoms was increasing. I began to panic as I felt my atoms getting farther and farther apart. My consciousness was flitting back and forth between my atoms as the space between them grew greater and greater. It felt as if I was desperately trying to keep them all together, to keep them in contact despite the increasing distance. I awoke to my body shaking all over, drenched in sweat.

In an effort to try to calm myself down, I tried to make some sense of this dream. I knew that it was just a dream and that I hadn't really left the bed, but it was still a terrifying experience. I racked my brain trying to understand why it was so frightening, and I finally realized that my dream was a warning not to go too far out, too fast. My desperate attempts to hold my body together as it expanded were a reflection of my terror of losing my integrity. Was I losing my bearings by talking about NDEs at a national professional conference? I was able to pull myself together, and delivered my presentation the next morning. But it was a much more humble and skeptical presentation than I'd been planning before that terrifying nightmare.

Five years into my position as a psychiatrist at Michigan, the chairman of the department called me into his office. I knew that my clinical work with patients was well respected and that my students gave positive evaluations of my teaching. I expected this to be a routine review of my performance. But my confidence changed to apprehension once I took a seat. Behind his immaculately ordered teak desk sat the balding senior professor, eying me above his half-glasses. I felt as if I were back in my home, about to be lectured by my loving but stern father.

With a restrained smile, he said that the medical school

was pleased with my clinical work and with my teaching, but what really counted toward promotion and tenure was research. He told me that I should stop wasting time studying NDEs because they were 'just anecdotes.' He said that in order to keep my job, I would need to carry out controlled laboratory experiments – where research participants are randomly assigned either to an experimental group or to a control group, and aren't told which group they are in. But of course, we can't assign people to have NDEs or not, and we can't keep them from knowing whether or not they've had NDEs. So any research into NDEs would not be considered when my contract came up for renewal – and might in fact be held against me.

I was devastated. My childhood fears of disappointing my demanding father flooded back in. Here was the department chairman, someone I looked up to as a valued mentor and ally, telling me I was not measuring up to his scientific standards, that studying NDEs was a waste of time. I struggled to keep my composure and said, 'That's not how I see near-death experiences.'

'Of course it's not!' he bellowed. 'That's why I'm telling you this. I know all about NDEs. My father had one, so I know how powerful they can be. But they're not something we can explain or study. If you keep wasting your time with this stuff, you won't be around here much longer.'

This was a major blow, and I wasn't sure how to handle it. Although most of my research had been focused on NDEs, I didn't think of myself primarily as an NDE researcher – nor did anyone else. I was first and foremost a clinical psychiatrist. I spent most of my time treating patients, and much of the rest of it teaching psychiatry to residents, interns, and medical students. I did research on NDEs primarily at night and on weekends. It was more or less a consuming hobby, since I wasn't getting paid to do it. Was my career as a psychiatrist at a medical school worth risking for the sake of this 'hobby'?

My chairman was trying to guide me to change my 'errant' ways so I could stay in his department. I knew I *could* do that. I could forget about NDEs and focus my research instead on mainstream psychiatric research into drugs and brain chemistry, and apply scientific methods to the mechanisms of mental illness. But I knew that NDEs really happened to people, and that they challenged our understanding of the mind and brain – and I couldn't pretend otherwise.

Whatever NDEs were, they were changing people's lives as surely as our psychiatric drugs and psychotherapy. What's more, they seemed to do this much faster, more profoundly, and more permanently. And beyond that, NDEs changed not only the experiencers' lives, but often the lives of others who came into contact with them – including me.

NDEs were things that I didn't understand, and things that no one else seemed to understand, either. But I wasn't getting paid to investigate them, and wasn't likely to.

Any time I was spending on this 'hobby' was time I was taking away from my family, which now included two young children. Family had always been the most important part of my life, as it was to my wife, Jenny. I was grateful for Jenny's support of my NDE research, but it always left me with mixed feelings. I had a very full and rewarding life without my research. I had my family, and I had my clinical and teaching job, both of which I cherished. Where did NDE research fit into the picture? There was no question that my life was family first and career second, NDE research third. How could I justify taking time from my nights and weekends with family and risking my career? Was studying NDEs that important?

The confrontation with my chairman brought all these issues to the fore, and I now had to face them head-on. If I wanted to keep my job, I would have to drop my interest in NDEs. But that just felt dishonest. The fact that NDEs were hard to explain or challenging to study was not, to my scientific training, a reason to abandon them. It was a reason to redouble our efforts to understand them.

But what about my chairman's accusation that NDEs are 'just anecdotes'? Researcher Arvin Gibson noted: 'The basic data for study *must* come from the stories of those who have undergone near-death experiences (NDEs), and to exclude the stories in an attempt to present some sanitized statistical version of the data would in itself be academically dishonest ... Without the stories, there would be no data to analyze.' I have thousands of NDE accounts in my files that are remarkably consistent, and I am just one of many scientists who have been studying these experiences over the past forty-five years. When so many people come forward separately relating similar personal experiences, it's worth taking a deeper look. In fact, throughout history, personal anecdotes have been the source of most scientific hypotheses.

Most research *starts* with scientists collecting, verifying, and comparing anecdotes until patterns in these stories become apparent, and then from those patterns emerge hypotheses, which can then be tested and refined. Collections of anecdotes, if they are investigated rigorously, are of immense value in medical research. They were critical, for example, in the discovery of AIDS and Lyme disease, and in discovering unexpected drug effects. As political scientist Raymond Wolfinger said a half century ago, 'The plural of anecdote is data.'

What would happen if we ignored anecdotes because they're not based on controlled laboratory experiments? If I tell my doctor I'm having chest pain and I'm having trouble breathing, I don't want to hear my doctor say, 'That's just an anecdote; it's not worth looking into.' I expect my doctor to take my symptoms seriously, as they could well be potential clues to something important.

What makes research scientific is a rigorous procedure for collecting and evaluating information. But that doesn't always involve controlled laboratory experiments, where research subjects are randomly assigned to an experimental group or a control group. Actually, very few topics of scientific research can be studied with controlled experiments. There are many fields that everyone accepts as science, even though laboratory experiments are difficult if not impossible – fields like astronomy, evolutionary biology, geology, and paleontology.

The prestigious *British Medical Journal* published a tongue-in-cheek article claiming to examine whether parachutes help prevent deaths in people who jump out of airplanes. The authors had eliminated anecdotal evidence from consideration, including in their review only randomized controlled trials. Of course, they couldn't find a single experiment in which people were randomly assigned to jump out of an airplane either with or without a parachute. They concluded: 'The perception that parachutes are a successful intervention is based largely on anecdotal evidence.' They went on to argue that scientists who consider only randomized controlled experiments would have to say that there is no evidence supporting the usefulness of parachutes! The authors did offer an alternative conclusion: 'That, under exceptional circumstances, common sense might be applied.'

Of course, it wouldn't make sense to take all anecdotes at face value without looking into them. And likewise, it doesn't make sense to reject all anecdotes without looking into them. I don't want my doctor to accept my chest pain at face value as proof that I'm having a heart attack, but I don't want him or her to dismiss my symptoms as meaningless anecdotes. I expect my doctor to look into my chest pain and evaluate it in the light of other evidence. The same is true of all anecdotes. It's equally unscientific to accept them all *or* to reject them all without evaluation.

Once I got over the blow of disappointing my chairman, I decided that understanding NDEs *was* important. But that would mean I was not going to be allowed to stay at Michigan, and I didn't want to stick around to be snubbed by a formal promotion and tenure review. I still loved treating patients and I loved teaching. After talking it over with my wife, Jenny, we decided I would apply for a job at another university, a clinically based medical school that would value my patient care and teaching enough to allow my pursuit of NDEs. That would mean uprooting my wife and kids, which was a lot to ask of them. I decided to look for a med school in the Northeast, near our two widowed mothers, our siblings, and their children. If I was going to move my wife and children across the country for my work, I wanted the move also to strengthen our family ties.

Our move to Connecticut turned out to be fulfilling both for my family and for my research. It felt like a home-coming, living close to our relatives again and working at a university that valued my patient care and teaching and allowed me to do any research that interested me, as long as it was done well. While there, I was lucky enough to have as research collaborators some people who themselves had had near-death experiences, and many who had not. As a scientist, I place tremendous value on intellect and crit-ical thinking, but I'm also aware that those assets can lure me into a one-sided view of the world. Being able to run my ideas by people who have actually done the 'field work,' who've had NDEs themselves, helps me see things from different perspectives and keeps me from getting side-tracked or stuck in academic blind alleys. And running my ideas by people who have less familiarity with NDEs repeatedly reminded me how startling these experiences can sound to those who haven't heard about them.

At that point in my career, even with the NDE Scale established as giving researchers around the globe confi-dence they were studying the same kind of experience, I knew I was missing important elements. Even as I was carefully building scientific tools and methods and using them for systematic research, I could see that there was more breadth and depth to NDEs than I could capture in a short-answer scale. Questionnaires can tell us many valuable things about NDEs, but they also miss a lot. There is a richness to the experiencers' own words that can't be captured in short-answer questionnaires. Experiencers were consistently telling me that, as helpful as the NDE Scale was in defining NDEs for research purposes, I would have to look deeper into the stories to get a better understanding.

6

....

Out of Their Bodies

Some aspects of NDEs, like the life review, seem easy for most people to relate to. Many of us review parts of our lives from time to time, especially in times of transition or major life events. But other NDE features are harder to understand. As science historian Thomas Kuhn pointed out, scientific advances often come about when a new fact is discovered that can't be explained. So I decided I had to dig deeper into those parts of the NDE that were hardest to explain. And as luck would have it, that's when Al Sullivan came into my life.

The fifty-six-year-old truck driver with a trim white beard showed up one evening at a support group I had started at the University of Connecticut for experiencers and others interested in NDEs. Al introduced himself to the group, but then sat silently through the entire meeting, looking attentive and occasionally smiling or nodding at something that was said. Toward the end of the meeting, I asked him whether, as a newcomer, he had anything he wanted to share. His eyes smiled as he said, 'Maybe next time.' As the group was disbanding, he approached me and asked if he could make an appointment to come see me in my office the next day.

When Al showed up for our appointment in his delivery uniform, he launched into his story without hesitation. 'I started having chest pains at work one Monday morning,' he began, 'and the dispatcher called the rescue squad. They took me straight to the hospital, and then while they were testing my heart to evaluate the problem, one of the main arteries to my heart became totally blocked.' He paused, but the smile never left his face.

'Ouch,' I said. 'What happened then?'

'Well, I don't remember it too clearly, because I was getting light-headed. But the surgeon said at least one of my coronary arteries was blocked and they'd have to operate immediately. I signed a piece of paper and asked them to call my wife and tell her. Then they rushed me to the operating room for what turned out to be quadruple coronary bypass surgery. Of course, I didn't know any of that. When I came to, I was looking down on the operating room from above.'

'That must have been a surprise,' I said.

'Well, that wasn't the real surprise,' Al continued. 'As I looked down, to my amazement, at the lower left-hand side was – of all things – me! I was lying on a table covered with light blue sheets and I was cut open so as to expose my chest cavity. In this cavity I was able to see my heart. I was able to see my surgeon, who just moments ago had explained to me what he was going to do during my operation. He appeared to be somewhat perplexed. I thought he was flapping his arms as if he was trying to fly.'

'What do you mean?' I asked. Al demonstrated by placing his palms on his chest and wiggling his elbows. This struck me as too bizarre to have really happened. A surgeon flapping his arms in the middle of an operation? In all my years in the medical field, I'd never seen or heard of a surgeon doing such a thing. It's also not the kind of thing you see surgeons do on television shows. It sounded a lot more like a strange dream caused by the general anesthesia Al had been given than like something he'd really seen his surgeon do.

I cocked my head and raised an eyebrow. 'Okay,' I said slowly, 'then what?'

'I know,' Al continued, 'it looked weird to me, too. But then I turned my attention to the lower right-hand side of the place I was at. Then, enveloped in warmth, joy, and peace, and a feeling of being loved, a brown-cloaked figure drifted out of the light toward me. As my euphoria rose, I recognized, much to my delight, it was my mother. My mother had died years ago when she was thirty-seven and I was only seven. I'm now in my fifties and the first thought that came to my mind was how young my mother looked. All at once my mother's expression changed to that of concern. At this point she left my side and drifted down toward my surgeon. She placed the surgeon's hand on the left side of my heart and then returned to me. I saw the surgeon making a sweeping motion as if to rid the area of a flying insect.' Al paused, and for the first time his smile left his face.

'And then?' I prompted.

'Well,' he said slowly, 'there's more, but I'm not sure I'm ready to talk about it just yet.'

'Oh?' I said, trying to think of a way to keep him talking. Before I could think of anything, he continued.

'Yeah,' he went on. 'I was told by some other being that a little boy who lives near me has cancer, and that I had to tell his parents.' He paused again and licked his lips. 'But I don't think I can do that to them. I mean, how could I explain to them how I know this?'

'I can see how that would be a problem,' I said. 'Why don't you think about that for a while? Maybe talk it over with someone.'

'There's no one I can talk to about this. My wife doesn't want to hear about it. She doesn't want to hear anything at all about my experience. She says she married a gentle, hardworking, funny guy, not an Old Testament prophet.'

'Maybe you could bring her to the NDE support group next month,' I suggested. 'Let her see that experiencers are just normal people, and that you're not the only one.'

Al laughed and shook his head. 'No,' he said, 'she'd never come. She doesn't even want *me* to go back. She says I think too much about my NDE and I need to forget about it and get back to reality.'

It was clear to me that Al had been through a profound experience that had changed his life, but I still couldn't accept at face value what he claimed to have seen in the operating room. I thought I might be able to put to rest his vision of the surgeon 'flapping' if I could speak with the doctor or someone on the operating team.

'Al, did you ever tell your surgeon what happened to you during the operation?' I asked.

'Oh, sure,' he said. 'Not right away, but several days later, when he came to my hospital room on his daily rounds. I asked him why he'd been flapping his arms in the operating room like he was trying to fly.'

'What did he say?'

'Well, he was embarrassed. He got angry, and asked, "Who told you that?" I said, "No one told me. I was watching you from up there," pointing my finger up toward the ceiling.'

'So what did he say about that?' I asked.

'He got very defensive,' Al said, 'like he thought I was accusing him of something. He said, "Well, I must have done *something* right, because you're still here, aren't you?" And then he walked out.'

Up to that point, I'd been identifying with Al, taking in his story from *his* perspective. But hearing his surgeon's reaction brought me back to Holly confronting me about the stain on my tie. I could readily identify with the surgeon's discomfort. It was more than just embarrassment. It was the seasick feeling of being involved in something that couldn't be real.

'Would you mind if I talked with him about it?' I asked.

'Be my guest,' Al said.

'I'll need you to sign a release form, because I don't work at his hospital.'

'Sure,' he said, 'go for it.'

Al's surgeon, a straitlaced Japanese American cardiac surgeon with an excellent reputation, did not seem to be someone prone to joking around in the operating room. He agreed to meet with me, and was eager to hear about Al's current condition. Much to my surprise, he confirmed what Al said. He told me that during his surgical training in Japan, he'd developed a peculiar habit that he'd never seen an American surgeon use. After he had 'scrubbed in' to the operating room and donned sterile gloves, he didn't want to risk touching anything in the room that might transfer a contaminant, however small, to his hands. So while he watched his assistants begin the operation, he planted his hands on his chest, flat against his sterile gown, to make sure he didn't accidentally touch anything. He then supervised his team, using his elbows instead of his fingers to point out various things.

Before that conversation, I'd suspected that Al's vision of the surgeon flapping his elbows had been a dream. But when I discovered that it had really happened, I had to search for another explanation. I asked the surgeon how *he* made sense of Al's claim to have seen all that. He shrugged. 'My family is Buddhist,' he said. 'Everything doesn't have to make sense to us.'

I began to wonder whether Al might have seen his surgeon 'flapping his arms' before he was completely anesthetized. So in order to pinpoint the time, I asked Al what else he had observed at the time his surgeon was flapping his arms.

He said that he saw his chest held open by metal clamps,

and two other surgeons working on his leg. That puzzled
him, because his problem was with his heart and he didn't
expect anyone to be messing with his leg. In fact, the surgeons were at that time stripping a vein out of his leg to be
used to create a bypass graft for his heart. That detail clearly
established that Al had been completely unconscious when
he'd witnessed the cardiac surgeon flapping his arms. He
couldn't possibly have seen that bizarre behavior with his
eyes, because his brain was fully anesthetized and his eyes
were taped shut – something that is often done to keep
patients' eyes from drying out if they're going to be anesthetized for a long time and unable to blink. He shouldn't
have been able to see anything. And yet he did.

As puzzling as Al's experience was, it was not unique. That
kind of accurate vision from an out-of-body viewpoint
doesn't happen very often in NDEs, but Al's wasn't the
only example I'd heard. Jane had an NDE at age twenty-
three while giving birth to her first child. She described
leaving her body and seeing things going on elsewhere:

'Due to blood loss, my blood pressure dropped. My
blood type wasn't available and nurses were in a panic.
When I heard a nurse say, "Oh, my God, we're losing her,"
I was out of my body at once and on the ceiling of the
operating room looking down, watching them work on a
body. I knew I wasn't dead. It took a while to recognize the
person I was viewing was *me*! I watched my doctor arrive
and procedures being done, heard conversations and saw
my baby being born. I also heard comments and concern

for her. It was a small hospital and I found myself over my mom in the waiting room. She was smoking. My mom doesn't smoke, but she admitted much later that she had "tried" one or two because she was so nervous! I returned to the operating room and my baby was doing better. I was not.'

Jane said she encountered her deceased grandmother and a 'guide' who told her that it wasn't her time and that despite her body having gone into shock, she had to return. She then awoke in her hospital bed with tubes in her arms. She tried to tell her nurses and her doctor, but they told her 'it was nothing,' and she realized it was something they wouldn't understand.

Colleen had a similar experience during an NDE at age twenty-two when she hemorrhaged after giving birth. She described for me her out-of-body perceptions:

'I was suffering intensely until I finally lost consciousness. Now, when I became conscious again, it was not at all in the normal way! In fact, not only was I out of my body, but it took me a few moments to realize that that cadaverously pale, blood-soaked body lying on the operating table was indeed mine! Let me say here that my "point of consciousness" was up somewhere near the ceiling. I was watching the bevy of nurses and doctors rushing madly around the room, all very much intent on bringing that poor young girl back to life.

'There was a tremendous heated discussion between my gynecologist and the anesthetist who had been summoned. My gynecologist was insisting that it was useless to try anything, because it was evident that it was much too late, that I was in fact dead, and that was it for him. I most definitely owe my life to the anesthetist, who fought to bring me back to life. I can still see him screaming, "She's just a kid. We've got to do something!" And he urged the nurses to do some transfusions and he literally forced the gynecologist to join the operating team. I remember being shocked at the swear words being used by the two doctors. I could not believe that doctors uttered such coarse things, and in the presence of nurses, too!

'When I regained consciousness – in the normal way, in my body – a few days later, I was in an intensive care unit, all hooked up to some IV machines, and a doctor came in the room. I recognized him immediately, and I thanked him for saving my life. It was the anesthetist. He seemed surprised that I should be doing so, and he asked me why I thought I should thank him for saving my life. So I recounted everything to him: how I had been present in the operating room and witnessed everything out of my body. I told him how shocked I'd been at hearing the coarse words exchanged between the gynecologist and himself. He was quite incredulous at first, but he urged me to tell him all else that I remembered. When I'd finished relating everything, he said that he was not overly astonished to hear my account, because he had been previously involved with a couple of other patients who had had near-death experiences.'

Colleen added that it meant a lot to her to have the anesthetist acknowledge that he'd heard other patients describe NDEs. She never doubted the reality of her experience. It had seemed to her more real than her every-day experiences. But it was important to her to be able to talk about it with a doctor who did not dismiss it as a hallucination or a dream. I heard comments like this again and again from experiencers. When healthcare workers discounted their NDEs, experiencers often became frustrated, angry, and depressed, and felt devalued as human beings. But when doctors and nurses listened to their stories – whatever they thought about NDEs – and acknowledged their importance to the experiencers, the patients felt respected and understood.

More than 80 percent of the experiencers who've participated in my research report having had a sense of being outside their physical bodies. However, only half of them describe actually seeing their bodies and observing events around them from a viewpoint above the scene, like Al, Jane, and Colleen did. Many experiencers are surprised to look down and see their bodies from a distance, and some don't even recognize it at first as theirs. Other experiencers recognize their bodies but are confused by being separated from them. Some experiencers have no idea they've died until they are surprised to come upon their lifeless bodies. Among the experiencers I've interviewed, this sense of leaving the body and later returning to it is usually described as easy, painless, and instantaneous.

Reports of seeing things from an out-of-body perspective during NDEs are not new. Sir Alexander Ogston, the noted Scottish military surgeon famous for discovering the

Staphylococcus bacteria, had an NDE when he was hospital-ized with typhoid fever during the Boer War in 1900, at the age of fifty-six. He described repeated out-of-body experiences during his NDE:

'Mind and body seemed to be dual, and to some extent separate. I was conscious of the body as an inert tumbled mass near a door; it belonged to me, but it was not *I*. I was conscious that my mental self used regularly to leave the body . . . until something produced a consciousness that the chilly mass, which I then recalled was my body, was being stirred as it lay by the door. I was then drawn rapidly back into it, joined it with disgust, and it became *I*, and was fed, spoken to, and cared for. When it was again left I seemed to wander off as before . . . and though I knew that death was hovering about, having no thought of religion nor dread of the end, and roamed on beneath the murky skies apathetic and contented until something again dis-turbed the body where it lay, when I was drawn back to it afresh, and entered it with ever-growing repulsion . . .

'In my wanderings there was a strange consciousness that I could see through the walls of the building, though I was aware that they [the walls] were there, and that everything was transparent to my senses. I saw plainly, for instance, a poor RAMC surgeon, of whose existence I had not known, and who was in quite another part of the hospital, grow very ill and scream and die; I saw them cover his corpse and carry him softly out on shoeless feet, quietly and surreptitiously, lest we should know that he had died . . . Afterwards, when I told these happenings to the sisters, they informed me that all this had happened just as I had fancied.'

Ogston's feeling of freedom from his physical body is echoed throughout modern accounts of seeming to leave the body in a severe medical crisis. Neuroanatomist Jill Bolte Taylor had a severe stroke that robbed her of her ability to walk, talk, read, write, or recall anything for years. When she finally recovered, she wrote a graphic account of her observations as the stroke gradually took over her brain:

'I remember that first day of the stroke with terrific bitter-sweetness . . . My perception of my physical boundaries was no longer limited to where my skin met air. I felt like a genie liberated from its bottle. The energy of my spirit seemed to flow like a great whale gliding through a sea of silent euphoria. Finer than the finest of pleasures we can experience as physical beings, this absence of physical boundary was one of glorious bliss. As my consciousness dwelled in a flow of sweet tranquility, it was obvious to me that I would never be able to squeeze the enormousness of my spirit back inside this tiny cellular matrix.'

Many of the out-of-body visions that experiencers describe are difficult for us to verify. Might they be just the experiencer's imagination, or lucky guesses about events that might have been expected to occur? At first glance, many accounts could seem to fall into either category. But there have been two studies that examined the accuracy of patients' descriptions of their own resuscitations when their hearts had stopped, comparing the accounts of patients who experienced NDEs with those who didn't.

Cardiologist Michael Sabom found that experiencers'

descriptions of their resuscitations were highly accurate, with very specific details of unexpected events. On the other hand, when he asked patients who were resuscitated but did not report NDEs to imagine what their resuscitations must have looked like, their descriptions were vague and contained many mistakes. Intensive care nurse Penny Sartori replicated Sabom's findings in a five-year study of hospitalized intensive care patients. Again, patients who reported leaving their bodies when their hearts stopped described their resuscitations accurately, but every cardiac arrest survivor who had not reported leaving the body made major mistakes in describing the equipment and procedures used.

How common is it for people who've had NDEs to describe accurately what happened around them while they were unconscious? Counseling professor Jan Holden reviewed ninety-three reports of out-of-body perceptions during NDEs. She found that 92 percent were completely accurate, 6 percent contained some error, and only 1 percent were completely erroneous. The fact that *any* out-of-body perceptions are accurate should be enough to make us scratch our heads. As William James, the father of American psychology, wrote more than a century ago, 'If you wish to upset the law that all crows are black, you mustn't seek to show that no crows are; it is enough if you prove one single crow to be white.'

The accuracy of out-of-body perceptions during NDEs makes it hard to dismiss near-death experiences as hallucinations. However, even when the reports are corroborated by independent witnesses – as when Al's surgeon acknowledged 'flapping his arms' – they are still anecdotal accounts reported after the event. If people can really see things from an out-of-body position while they're unconscious, shouldn't we be able to test that with a controlled experiment?

Since 1990, there have actually been six published attempts to test the accuracy of out-of-body perceptions during NDEs. In these experiments, researchers place unexpected visual 'targets' in locations where they might be seen by people having NDEs. Researchers generally place these targets in the upper corner of patient rooms in the emergency department, in the coronary care unit, and in the intensive care unit of a hospital – areas where there's a good chance that patients' hearts might stop. Patients are not told about the targets, but any who claim to have left their bodies are asked whether they saw anything unusual or unexpected in the room.

Altogether, these six studies found a total of only twelve patients who reported NDEs and felt they'd left their bodies. None of those twelve patients reported having seen the visual target, leaving researchers with no evidence bearing on the question of whether patients having NDEs can or can't see from an out-of-body perspective.

I'd grown up watching doctor shows on television in which almost every patient whose heart stopped is successfully brought back to life. Thus I was surprised to learn, when I got to medical school, that surviving a cardiac arrest is rare, even in the hospital. A 2018 update from the American Heart Association found that the overall survival rate for cardiac arrest was only 10 percent outside the hospital and 25 percent in the hospital – and most of those resuscitations are only for brief periods. Only 11 percent survive long enough to go home from the hospital.

With this difficulty in mind, I designed a study with patients who I *knew* would survive the cardiac arrest. I carried out an experiment with patients whose hearts were being stopped in a carefully monitored and controlled situation. These were patients who have repeated dangerous heart arrhythmias and are at high risk of their hearts suddenly stopping. Such patients are often referred to a heart surgery clinic to have a small device surgically placed in their chests. This device, called an implantable cardioverter-defibrillator, or ICD, continually monitors the patient's heart rhythm. If it detects that the heart has stopped, it will automatically shock it back into a normal rhythm. When the cardiac surgeons place the device inside the patient's chest, they have to test whether it will work, before they close the chest up again. They do that by intentionally stopping the patient's heart with an electric shock, and then waiting to see whether the ICD will restart the heart again. Since we know exactly when and where the heart will be stopped, we know when and where to place a visual target for the patients to see, if they do leave their bodies.

Cathy Milne, a cardiac clinic nurse, had studied the frequency of NDEs in people undergoing this procedure ten years earlier. She found that 14 percent of them reported NDEs. So I had good reason to think that we'd find enough near-death experiences to test the patients' ability to see a visual target. I worked with Jan Holden to plan the details of the study and of the targets. I placed a laptop computer on top of a fluoroscopy monitor high above the operating table. The computer was programmed to select at random one out of seventy-two unexpected animated images, such as a purple frog jumping around the computer screen. The animation would run for five minutes, interrupted by a flashing time display, and then shut itself off. If patients actually left their bodies when their hearts were stopped, they might be able to identify the image. The computer kept a record of what the target was for each patient, but none of the staff in the room would know the target.

After each patient woke from the anesthesia, I asked the usual question I had used in my study of patients hospitalized after a cardiac arrest: 'What was the last thing you remember before you blacked out?'

Half of these patients, however, gave me a puzzled look and said, 'What do you mean? I didn't black out.'

As it turned out, I hadn't appreciated the effect of the new sedative, midazolam, that was routinely given to each patient before the procedure, in order to make them drowsy and reduce their anxiety. Midazolam is used because it usually stops patients from remembering the procedure. That's helpful when the goal is to reduce any memory of the painful shock to their hearts. But it's not helpful when the goal is to explore any memories of having left their bodies when their hearts were stopped. That is, it makes the procedure easier for the patient, but makes it harder to remember any experience the patient may have had. After more than fifty induced cardiac arrests in this study, I didn't find a single patient who remembered anything remotely like an NDE or leaving the body during the procedure.

Beyond the difficulty remembering with midazolam, there was another problem with my experiment that I hadn't anticipated. When I discussed these research findings at a conference attended by a large number of people who'd had NDEs, they were astounded at what they considered my naivete in carrying out this study. Why, they argued, would patients whose hearts had just stopped and who are being resuscitated – patients who were stunned by their unexpected separation from their bodies – go looking around the hospital room for a hidden image that has no relevance to them, but that some researcher had designated as the 'target'?

I had high hopes that these studies would either confirm or disprove experiencers' claims that they could see accurately during their NDEs. The failure of this line of research to provide a definitive answer was a disappointment, to put it mildly. The skeptic in me kept saying that unless we could demonstrate accurate out-of-body perceptions during NDEs in a controlled experiment, we couldn't know whether or not we were fooling ourselves. Could Al Sullivan have overheard some nurses talking about his surgeon's peculiar arm-flapping habit, and then imagined having seen it himself? Could Holly's report of the stain on my tie have been just a coincidence or a lucky guess? After all, it was 'only an anecdote' and not the result of a controlled experiment. Was her correct report of my tomato-sauce stain less impressive because the stain was not intentionally planted? Do you doubt that parachutes save lives because the evidence comes from anecdotes rather than from randomized experiments?

Of the roughly seven hundred experiencers in my study of out-of-body vision, four out of ten reported becoming aware of events beyond the range of their senses. Of those, half said that they later checked out their perceptions with other people, who confirmed that the events they 'saw' or 'heard' really happened. But the other half never told the people involved in these perceptions – usually doctors and nurses – because they were afraid it would sound too outlandish. I could easily understand how experiencers would be afraid that other people – particularly doctors – might think they were 'crazy' if they talked about these extraordinary perceptions. I remembered Bill Hernlund, the firefighter who was caught in an airplane explosion, who had told his doctor that he had died and come back and was subsequently scheduled for a psychological evaluation.

These experiencers' stories about seeing and hearing things while they were unconscious made me question my childhood belief that nothing exists except what can be seen, heard, or felt. Certainly Holly's account of the stain on my tie had challenged my beliefs, but that one incident alone didn't convince me of anything. Although I couldn't deny it had happened, I continued to wonder whether Holly could have gotten the information in some 'normal' way that I couldn't identify. Many scientists – including me, up to a certain point – try to ignore these unexplained events, and when they become troubling enough, try to deny they ever happened. But as neuroscientist and anthropologist Charles Whitehead wrote, 'Anomalies tend to get swept under the carpet until there are so many of them that the furniture starts to fall over.' The sheer volume of these unexplained incidents in which rational scientists were stumped by experiencers seeing and hearing things that they couldn't have seen or heard – like my incident with Holly, and like Al's with his surgeon flapping his arms, and the hundred others documented by Jan Holden – were beginning to topple the furniture in my worldview.

For most people – as they did for me – these more extraordinary features of NDEs straddle the 'boggle threshold,' the point in an otherwise true story at which it becomes so outlandish that you begin to question its believability. In some cases, the events they described could be verified. But even in cases where there were no independent witnesses, the experiencers I'd interviewed were uniformly sincere and so profoundly affected by their NDEs that I couldn't imagine them lying about their experiences. Their NDEs needed to be treated with respect and taken seriously. But looking down from the ceiling at your body lying on an operating table is difficult to explain, and I knew I needed to come up with some working theories about what could be happening. One theory often suggested by people seeking to dismiss NDEs is that they're fantasies that didn't really happen outside the experiencer's imagination. As a psychiatrist, I knew I'd have to address that question head-on.

Or Out of Their Minds?

The fall from the roof of his college dorm broke both of Peter's legs. He was an undergraduate who was being treated at the student health clinic for hearing voices. I interviewed him a few days later in his hospital bed, as he was recovering on the orthopedic ward. His psychiatric medications had been restarted, and, according to the nurses' notes, he was no longer hearing voices or feeling confused, as he had been when he was admitted to the hospital.

'Peter,' I began, after introducing myself, 'I understand that you jumped off the roof of your dorm. Can you tell me about that?'

He took a deep breath and launched into his story. 'I'd stopped my meds because they were making me tired all the time. It was getting hard for me to concentrate on my

studies. After a couple of weeks, I started hallucinating.'
He paused and looked at me.

'Hallucinating?' I asked.

'Yeah, I was hearing the Devil tell me that I was now his, because I'd messed up so much.' Peter was looking down at his legs in casts, and picking at the sheet. 'He said that I was evil and that I had to die, and he was calling me to join him in hell. So I went up the stairs to the top floor of my dorm and climbed the ladder to the roof. I sat on the parapet at the edge, with my legs dangling over the side. It was early morning, and no one was walking down below. I started shivering with the cold, and Satan started yelling at me, "Do it now! Do it now!"

'I was confused and I was terrified,' he continued. 'The hallucination wouldn't stop, and I believed what it was saying. I thought I didn't deserve to live, and so I eased myself forward and pushed off with both hands. I closed my eyes and felt sick to my stomach as I fell.'

Peter looked up at me. After a brief pause, as if considering how much to tell me, he picked up the story. 'But then, as I was falling, God spoke to me. I couldn't see him, but I heard his strong, clear voice. He said, "Peter, you are one of my children. You do *not* belong to Satan. You are loved more than you will ever know. I will not let your life end like this."'

Peter stopped and licked his dry lips. He reached over to the table beside his bed and took a sip of water through a straw. He didn't continue his story.

'And then?' I asked.

'I don't know if I blacked out when I hit the ground. It

seemed like I lay there a long time, and it's all confused in my mind. I remember a crowd around me, and then being moved onto a stretcher and then into an ambulance. I was in a lot of pain, and I didn't fully understand what was happening. But I was relieved that I was still alive.'

'How are you feeling now?' I asked.

'Well, I don't want to kill myself, if that's what you're asking. I know that God has a plan for me.'

I nodded, trying to formulate the next question in my mind. 'You're telling me that you heard a voice that you identified as Satan, but you called that a hallucination. And then you heard another voice that you identified as God. Both of them were voices that only you could hear.'

'Yeah,' he said, nodding with a hint of a smile. 'I know what you're thinking. Why do I think one was a hallucination and the other was real?'

'Exactly,' I said. 'As I'm listening to your story, I have no way of telling those two voices apart. How do you?'

Peter shook his head slowly. 'I don't know how to explain it to you. But God's voice was stronger and clearer and more real to me than your voice is right now, just like your voice is stronger and clearer and more real to me than the Devil's. I know that leading up to my suicide attempt, I thought Satan's voice was real, but now that I'm no longer crazy, I know that it was just a hallucination. But not God's voice. That was real.' He waved his hand in the air. 'It was realer than all of this.'

Was hearing the voice of God as he was in midair part of Peter's illness, or was it an NDE? Although restarting his medications convinced him that Satan's voice had been a hallucination, it did not shake his belief that God had really spoken to him.

This distinction cut to the heart of my dilemma. The two voices that Peter heard – God's and Satan's – were quite different to him. But how can the rest of us know which voices are imaginary and which – if any – are real? It seemed to me that exploring the differences was key to understanding NDEs. As a psychiatrist, I was in a better position than some other NDE researchers to explore symptoms of mental illness in people who had NDEs. Furthermore, my clinical role at the university hospital put me in a position to treat a number of psychiatric patients who also had NDEs. The problem was to sort out the two: Did the mental illness contribute to the NDE? Did the NDE contribute to the mental illness? Were the two completely separate things that had nothing to do with each other?

My first approach in addressing these questions was to look at how common mental illnesses are in people who have NDEs. Do experiencers have any more or any less mental illness than non-experiencers? To answer this question, I compared the frequency of different psychiatric conditions among two groups of people in the general population: those who had reported NDEs, and those who had come close to death but had *not* reported NDEs. I used the Psychiatric Diagnostic Screening Questionnaire, a standard questionnaire used to screen for the sixteen most common psychiatric disorders, including depression, anxiety, post-traumatic stress, obsessive thoughts and compulsive behavior, eating disorders, alcohol and drug abuse, physical symptoms caused by emotional distress, and difficulty telling what's real from what's imaginary.

I found that each one of the sixteen specific conditions occurred at the same rate in people who'd had NDEs and in people who had come close to death but hadn't had NDEs. These conditions also occurred among experiencers as often as they do among the general public who *haven't* come close to death. I also compared these rates with the rates of these mental illnesses in the general population as a whole – and found no difference. In other words, the evidence suggests that having an NDE doesn't make people any more or less likely to have mental illness.

I also looked deeper into the rates of two particular illnesses that I thought might be connected to NDEs: *dissociation* and *post-traumatic stress disorder*. Dissociation is a condition in which your sense of self becomes detached from your bodily sensations. A normal and fairly common example is 'highway hypnosis,' in which you can drive sometimes for long distances, while your body responds appropriately to the road and to cars around you, but you are not consciously aware of driving until you suddenly 'come to' and maybe realize you've passed your exit. In extreme dissociation, you may feel you are no longer in your physical body. Dissociation can be a normal reaction to trauma, in which you protect yourself by walling yourself off from feeling the pain and fear of what's happening to your body.

I gave a standard measure of dissociation, the Dissociative Experiences Scale, to a group of people in the general public who'd come close to death. Some of them had NDEs during the brush with death, and some didn't. Those who had NDEs did in fact describe more features of dissociation, but far fewer than the amount required for a psychiatric diagnosis of dissociation. The degree of dissociation shown by experiencers was typical of a general response to trauma, but not typical of a mental illness. In other words, the experiencers shifted their attention away from the physical body in crisis. That shifting of attention is a normal response to intolerable trauma, and not a sign of mental illness.

I also wondered whether NDEs might be associated with post-traumatic stress disorder, or PTSD. It seemed inescapable that coming close to death would be pretty scary for most people, with the risk of dying, the pain, and the loss of control. It made sense that coming close to death might lead to PTSD, whether or not it was accompanied by an NDE. The characteristic symptoms of PTSD include vivid dreams or flashbacks of the crisis and efforts to avoid or block out reminders of the traumatic event. My father had a heart attack at age thirty while driving to work on a Monday morning, and after he recovered he was afraid to drive again. He saw a psychotherapist in order to overcome his fear of having another heart attack while driving. That was long before PTSD was an official diagnosis, but that's what his problem would be called now.

As I did with dissociation, I looked at people in the general population who had come close to death. I gave them a standard measure of PTSD, the Impact of Event Scale. I compared features of PTSD in those who had

NDEs and those who didn't. As with dissociation, I found that those people who had NDEs also had more features of PTSD, but far fewer than the amount required for a diagnosis of PTSD. Unlike people who have PTSD, those who had NDEs tended to have dreams and flashbacks of the brush with death, but no attempts to avoid reminders of it. That fits with what experiencers usually say about their NDEs – that the experience becomes a central focus of their life, but they don't think of it as a negative event to avoid. That particular pattern – flashbacks of the event without avoidance of reminders – is typical of people struggling to understand an experience and fit it into their lives, but it is *not* typical of a mental disorder.

So the evidence produced by a variety of studies suggests that experiencers do not have any more or any less mental illness than the general population. And specifically they do not have more or less dissociation or PTSD – illnesses that might be expected after a close brush with death.

Having answered the question of the rate of mental illness in people who've had NDEs, I also wanted to look at the other question: What is the rate of NDEs in people with mental illness? Do people who seek psychiatric help have more or fewer NDEs than the general population?

To answer that question, I turned to people seeking psychiatric treatment at my hospital's outpatient clinic – people who were doing well enough not to need hospital care, but still suffering from emotional distress. During the one year I carried out that study, I surveyed a total of more than eight hundred patients. As part of their routine initial interview at the clinic, I gave them a standard measure of psychological distress, the revised ninety-item Symptom Checklist, or SCL-90-R. I also asked them whether they had ever come close to death, and those patients who said they had also completed my NDE Scale.

One-third of the patients seeking psychiatric help said they had come close to death, and about 20 percent of those who'd come close to death also had NDEs. That's about as often as NDEs occur among people in the general public who come close to death. In other words, the evidence suggests that people with mental illness have NDEs no more or less often than do people in the general population.

In that outpatient clinic study, patients who had come close to death did have higher scores on the SCL-90-R than those who hadn't come close to death. That is, those who'd almost died experienced more distress than other patients. That didn't surprise me, because coming close to death is a traumatic event that often causes psychological distress. But what I *did* find surprising was that, among those who'd come close to death, those who'd had NDEs described *less* psychological distress than those who *hadn't* had NDEs. In other words, the evidence suggested that NDEs in fact offer some *protection against* psychological distress after a close brush with death.

So I did not find any link between mental illness and NDEs. People in the general population who have NDEs have the same amount of mental illness as do others. And people with mental illness have the same rate of NDEs as do other people. And what was possibly really good news is that NDEs may help protect against more severe psychological distress from a close brush with death.

But what about those people who have *both* NDEs and mental illness, like Peter? Can we tell the difference between his NDE and his illness in his experience when he jumped off the roof? This question comes up not only with mental illness but also with drug intoxication. When people who overdose on drugs claim they had an NDE, how do we separate the visions of an NDE from drug-induced hallucinations?

Justin overdosed on LSD at age eighteen at a college party. He collapsed on the floor and appeared to stop breathing. He described his consciousness as 'crystal clear' in his NDE, in contrast to the terrifying confusion of his LSD trip. Here's how he described the difference:

'My father had died of cancer a year earlier. I had gone to college because I felt programmed to do that, but I had no direction in life. One evening, a dorm friend invited me to his friend's home to try LSD. I was given three tablets, and wondered whether that was too much, but it seemed to this guy that he knew what he was doing. After about forty-five minutes of smoking hashish, I began hallucinating wildly. I hung on, trying to stay with it, but I felt like I was on a roller coaster. My mind was out of control, so helpless, and it felt like it was destroying me. I was getting more and more frantic and more and more depressed. I wanted out, but couldn't stop it. My worst nightmares were coming true. Then at some point I fell face-first down onto the floor. My friend later said I stopped breathing. Darkness swept over me, and I knew I was going to die and there was nothing I could do about it.

'The next thing that happened was a complete separation of my consciousness from my body. My consciousness, which I never lost, remained completely intact. I had no thoughts about my body lying on the floor. I was definitely not suffering anymore. I had been through hell with the LSD, but as soon as I left my body, that hell left me. I was not in pain anymore, and I was engulfed by the purest, most unselfish, most beautiful love I've ever felt in my life. I had absolutely nothing to do with the room I had collapsed in. I was 100 percent involved in this experience. I felt clear and "on my toes" throughout this experience.

'One thing must be made clear here: The clarity and easy perception of this experience is in direct antithesis to the anguish of the drug overdose. The drug trip was just terrible, absolute insanity at its lowest. More than anything, I wanted to snap out of it and return to normal. I desperately needed medical attention. After falling down, face-first on the floor, that's when I crossed over to the other side and left my body behind. At the point of freedom from suffering from the drug trip, everything became very clear. The NDE was crystal clear, as in waking up and rising to a new day. But, when I later woke up during the night at the hospital, I was once again hallucinating and very groggy from LSD, and probably from the hospital's treatment.

'That, to me over the past years, has been the wildest thing to come to grips with: an "ABA" effect, "A" being the bad drug experience, "B" being the crystal clear near-death experience, and again "A," waking up and still hallucinating and being unaware still of the difference between reality and that which was unreal. The NDE was a substantial, real, perfectly vivid experience, much more so than anything preceding it during the part of the LSD trip before or after the NDE. This fascinating personal experience has never diminished in intensity in the fifteen years since it happened.'

Justin drew a definite distinction between his clear consciousness in his NDE and the terrifying confusion of his drug trip, just as Peter contrasted the reality of his NDE from the unreal quality of his schizophrenic hallucinations.

And so did Stephen, a twenty-five-year-old nurse who had an NDE when he decided to end his life by taking a suicidal overdose of opioids. I approached Stephen in his hospital room just as I did Peter.

'Stephen,' I began, after introducing myself, 'I understand that you took an overdose yesterday. Can you tell me about that?'

Stephen muted the television mounted on the wall and looked me over. 'I'm in big trouble,' he said finally.

'Oh?'

He sighed, glanced at the doorway, and then launched into his story. 'I'd been taking opiates from the patients' supplies on the ward. Just a little at first, mostly oxycodone.

I only took it on shifts where someone else was doing meds, so the pill counts on my shifts would be accurate. But then I started taking more and more, and I'm pretty sure my manager was onto me.'

He paused, so I prompted him to continue. 'And so . . . ?'

Stephen took a deep breath and went on. 'I'd been feeling pretty stressed anyway. My dad had died a few months ago, and my girlfriend made it pretty clear she wanted to break up. I guess that's why I started taking the opiates. They were helping me relax, or maybe just blot everything out.'

'But they didn't solve your problems?' I asked.

He laughed. 'I knew they wouldn't. I was just trying to buy myself some time, until I felt strong enough to deal with everything that was going on.'

Stephen took off his glasses, wiped them with a tissue, and put them back on. 'But when my manager figured out that I was probably responsible for the repeated low opioid counts, I knew I wouldn't be able to get out of that one. I thought about just running, but I knew I'd have to face this sooner or later.'

'So your manager never actually confronted you?'

'No, but I knew she'd figured it out. It was just a matter of time before I'd be arrested. I decided I'd rather just end it all.' He looked down and shook his head. 'I stole a large enough number of pills to do myself in, and left my shift early, before anyone could do a pill count.'

'Sounds like you felt you'd painted yourself into a corner,' I suggested.

'You could say that,' he agreed. 'I went right home, took all the pills with a bottle of beer, and lay down on my bed, expecting that would be it.'

'And what were you hoping would happen next?' I asked.

'Nothing,' he said, quickly, as if he was surprised at the question. 'I thought I'd just drift off and that would be the end. Sooner or later, someone would come looking for me and find me. But it would be too late.'

'And what did you think would happen to *you* then?' I asked.

Stephen looked up at me, puzzled at first, and then amused. 'What, you mean like would I be judged and sent to hell? I never believed in anything like that. When you die, you die.'

He shook his head, but didn't say anything else.

'But you didn't die,' I said. 'What happened?'

'I think I might have dozed off, but then I awoke, and I had incredible cramping. I felt really nauseous, like I was going to vomit. And it was hard to breathe. I didn't seem to have the strength to take a deep breath. I was afraid I *wasn't* going to die. I was afraid I'd just give myself a stroke or something. And the stomach cramps were really bad. I thought I'd better get some help before I ended up in worse shape than I already was. The telephone was on the wall in the kitchen, maybe thirty feet from my bed. I tried to get up, but I was so dizzy I had trouble standing. I held on to the bed to steady myself, and then took a few steps toward the kitchen. I was real groggy, and it was hard even to stand up, let alone walk.'

He stopped, and after a few moments I encouraged him to continue. 'And then?'

He stared at me, and then after a long pause, continued his account: 'I was also hallucinating. As I stood there, swaying, steadying myself with a hand on the wall, I saw all these midgets in my apartment milling around my legs, making it harder for me to walk.'

'Midgets?' I wasn't sure I'd heard him correctly.

'Yeah,' he said, 'small people, about this high.' He held his hand out, palm down, about even with the bed. 'I know it sounds crazy. It *was* crazy. I was seeing things. But at the time, they seemed very real.'

Stephen swallowed, and then continued. 'It was very confusing. But then, all of a sudden, I felt myself leave my body.'

'You left your body?' I repeated, not certain I'd heard him correctly.

'Yeah. Well, I'm not sure I actually felt myself leave, but there I was, maybe ten feet behind my body, and looking down on it.'

He continued to stare at me, and I prompted him again. 'And what was your body doing?'

Stephen shook his head. 'It was just standing there, with my hand braced against the wall, looking down at all these midgets, trying to figure out what they were doing there.'

I didn't know what to say, so I just spread my hands, palms up, and raised an eyebrow.

'I mean, *I* couldn't actually see them,' he continued. 'Up there near the ceiling, I was watching my body sway-ing and looking down between my legs. I knew my *body* was looking at the midgets all around me, because I remembered being in my body seeing them. But from where *I* was . . .' He shook his head and swallowed. 'From my viewpoint looking down at my body, I couldn't see them. I knew that my body was confused and hallucinating, but my mind was crystal clear. *I* wasn't hallucinating, but my *body* was. I was thinking as clearly as ever, but my body looked totally dazed.'

He stopped again, and after a while I said, 'That's quite a story. How do you make sense of it?'

He snickered and shook his head. 'Beats the heck out of me! One minute I was there in my body, seeing midgets, and the next, I'm up near the ceiling. I have no idea what happened.'

After another long silence, I asked: 'And then?'

Stephen sighed. 'I think I must have blacked out again. I woke up lying on the floor, still pretty woozy, but the midgets were gone. I crawled into the kitchen, made it to the phone, and called the rescue squad.'

How could *I* make sense of Stephen's claim that he was not hallucinating, while his brain was? It seemed similar to Justin's distinction between the terrifying confusion of his LSD trip and the peaceful clarity of his NDE – and similar to Peter's distinction between the hallucinated voice of Satan and what he insisted was the real voice of God as he was falling off the roof.

While some unusual experiences that involve visions and unconventional beliefs may be due to mental illness, might other unusual experiences that involve visions and unconventional beliefs be authentic experiences – as Peter, Justin, and Stephen maintained? If so, how can we tell them apart?

What is the difference between mental illness and NDEs? I had to look for the answer in what happens *after* the experience itself, and take into account the role the experience plays in the person's life. Psychiatrist Mitch Liester and I compared two groups of people who reported repeatedly hearing voices. We compared people with schizophrenia and the small number of people who'd had NDEs and continued to hear voices after the experience. We asked both groups a series of standard questions about how helpful or harmful the voices were.

We found striking differences between the two groups. Most of the people who'd had NDEs found the voices were soothing or comforting, made them feel better about themselves, and had a positive impact on their relationships with other people. On the other hand, most of the schizophrenics found the voices were distressing or threatening, made them feel worse about themselves, and had a negative impact on their relationships. Most of the people who'd had NDEs wanted to keep hearing the voices, while almost none of the schizophrenics did.

So hearing voices that no one else can hear was a very positive experience for people who'd had NDEs but a very negative experience for the schizophrenics. Peter's experience of hearing the voice of God as he was falling might have sounded like one of his schizophrenic hallucinations, but it helped him find a sense of meaning and purpose in life – a clear and important distinction, as I was learning.

A number of clinicians have written about the distinction between unusual experiences that are symptoms of mental illness and those like NDEs that are spiritually transformative. One difference is that NDEs are triggered by life-threatening or otherwise extreme events. Also, they are usually short-lived and happen only once, and they often happen to people who are leading normal, productive lives. Mental illness, on the other hand, can happen without any obvious trigger, tends to go on for long periods or to recur again and again, and often happens to people with significant psychological difficulty or marginal social functioning.

Another difference is that NDEs and episodes of mental illness vary greatly in how they are remembered later on. NDEs are remembered vividly for decades after the event, and are often remembered as 'realer than real.' Their memory doesn't fade over time, but retains its vividness and richness of detail. In contrast, people with mental illness usually realize, after the acute episode is over, that their visions were unreal. And memories of an episode of mental illness fade over time, becoming less vivid and less detailed until they are completely forgotten, as are most dreams.

Additionally, people who have had NDEs often reexamine their experience again and again, in order to seek and develop insight into the meaning of the experience. They often seek out other experiencers to share their NDEs and insights. If they're disturbed at all by their NDE, it's usually only until they can figure out how to weave the experience and its lessons into their lives. On the other hand, people with mental illness usually avoid any reminders of their unusual thoughts and perceptions and don't seek to understand them. They generally don't want to share their experiences, which they most often find permanently disturbing.

Finally, NDEs usually lead to an enhanced sense of meaning and purpose in life, increased joy in everyday things, decreased fear of death, and a greater sense of the interconnectedness among all people. As a result, people who have NDEs often become less absorbed in their own personal needs and concerns and more altruistic and compassionate toward others. They often have positive outcomes after their NDEs, and don't generally struggle with day-to-day life. On the other hand, people with mental illness may lose their sense of meaning in life and joy in daily activities, feel more fearful and isolated from other people, and become more absorbed in their own needs and concerns and less involved with others. Mental illnesses often lead to negative outcomes, including difficulty maintaining jobs and relationships, legal complications, and struggles with harmful impulses.

Of course, all these distinctions between NDEs and mental illness are generalizations, and there will be exceptions to them all. Certainly some people do learn and grow from their mental illnesses. And some people may struggle for years to try to understand and integrate their NDEs into their lives. But those are the exceptions.

*

So the evidence suggests that NDEs are not associated with mental illness. I felt relief that this question had been put to rest. Over the years, I'd heard from too many people, like Bill Hernlund, that they'd been referred for psychiatric help because they'd shared their NDEs with a doctor who'd taken that as a sign of mental illness. I have been sharing this information at professional conferences and in my clinical supervision of medical students, residents, and chaplains at the hospitals where I've worked. And I've seen this information begin to influence the practice of healthcare. In recent years, it has encouraged new approaches to patients, as healthcare providers have become more sensitive to the frequency and normality of NDEs among their patients.

But if NDEs are not hallucinations or other phenomena associated with mental illness, does that mean they're real experiences? I needed stronger evidence to answer that question.

Are Near-Death Experiences Real?

The evidence seems to show that near-death experiences are very different from hallucinations. But that by itself doesn't necessarily mean NDE accounts are accurate descriptions of events. From time to time over the past four decades, I continued to wonder whether NDE narratives are memories of experiences that really happened, or are reflections of dying people's hopes and expectations. Some of my medical colleagues dismiss near-death experiences as pure fantasy and for that reason regard any research on NDEs as unscientific. But what makes an investigation scientific is not the topic being studied. What makes an investigation scientific is whether it's based on rigorous observations, on evidence, and on sound reasoning.

As neuroscientist Mark Leary wrote, 'Science is not defined by the topics it studies but rather by its approach to investigating those topics . . . The fact that some people do not believe that a phenomenon is real does not make research on that phenomenon pseudoscientific. Science can be used to address a wide array of questions, even questions about phenomena that ultimately turn out not to exist. In fact, one important function of science is to demonstrate empirically which effects are real and which are not . . . so it makes no sense to assert in advance that a study of a particular topic is not scientific because the hypothesis being tested is false!'

There are many examples throughout history of things rejected as unreal and therefore unworthy of scientific study, which were later shown to be quite real. Most scientists thought reports of meteorites were tall tales not worth investigating until the nineteenth century, even though there were accounts going back to ancient times of rocks falling from the sky. And scientists and physicians ridiculed the idea of germs until the nineteenth century, even though the ancient Greeks speculated about disease spreading from infected patients by 'seeds of plague' invisible to the naked eye. As recently as the 1980s, most medical scientists thought that looking for bacteria that might cause stomach ulcers was a waste of time – even though that idea is widely accepted today and earned Barry Marshall and Robin Warren the Nobel Prize in Medicine in 2005.

Some of my colleagues argue that NDEs can't be real because they contradict our current beliefs about how the brain works. But science by its very nature is always a work in progress. Each generation of scientists looks back on the models of previous generations with amusement at their naivete. Why then should we expect any of our current scientific views of how the brain works to stand up to the scrutiny of future generations?

Refining our models when new phenomena are discovered is how science makes progress. A century ago, advances in technology allowed physicists to explore new phenomena involving very small particles and very fast speeds. The formulas physicists had been using for centuries – formulas that worked very well in describing physical movements in our everyday world – weren't as accurate with these new phenomena. Physicists, to maintain their scientific integrity, could not ignore these new phenomena that didn't obey Newton's classical laws of motion. But that didn't mean they had to discard the old formulas as useless. They just had to acknowledge that Newton's laws were useful *only under certain conditions.* They had to refine the old formulas by blending mathematical calculations from relativity theory and quantum mechanics with classical physics to come up with a more complete model of reality.

In the same way, advances in medical technology in the past century have allowed neuroscientists to learn about near-death experiences and other phenomena involving continued consciousness when the brain is impaired. The model medical scientists had been using for centuries – a model that worked very well in describing the nonphysical mind as the product of the physical brain – worked in everyday life but couldn't account for these near-death experiences. Neuroscientists, to maintain their scientific integrity, cannot ignore NDEs that don't seem to fit into the old model of brain and mind. But that doesn't mean they have to discard the old model of the nonphysical mind being produced by the physical brain. They just have to acknowledge that the old brain-mind model is useful *only under certain conditions*. They must refine their old model to accommodate things like NDEs, where consciousness continues after the brain has shut down, in order to come up with a more complete description of reality.

Scientists never have final answers. What we have are observations, from which we weave stories that make sense of the evidence. In telling these stories, we have to keep them logically coherent and consistent with all the empirical observations. The result of all this repeated storytelling is that science is always marching toward an end it can never reach – a complete description of reality. As neuroscientist Thomas Schofield put it, 'Science is not about finding the truth at all, but about finding better ways of being wrong . . . A theory can never be perfect: the best it can be is better than the theory that went before.'

Astrophysicist Neil deGrasse Tyson makes the distinction between personal truth, which may be convincing to you as an individual but which you can't necessarily prove to anyone else, and objective truth. 'Objective truth,' Tyson says, 'is the kind of truth science discovers. And it's the kind of truth that is true whether or not you believe in it. It exists outside of your culture, your religion, your political affiliation.'

As I look deeper and deeper into near-death experiences, it seems to me that the research findings on NDEs meet Tyson's criteria for the kind of objective truth that science discovers. People from different cultures and different religions have NDEs, *whether or not they believe in them*. Some of the experiencers I've quoted in this book describe things that contradict their cultural and religious beliefs. Some were atheists who didn't believe in a higher power or in anything after death – but who can't deny their experience of somehow being conscious when their bodies were declared dead. It seems clear to me that the study of near-death experiences can be a rigorous, empirical, observational science.

Of course, in dealing with observations, we need to take into account the biases – conscious or unconscious – of the people collecting the NDE accounts. As in all research, we constantly have to monitor our own biases and question how they may be influencing our interpretation of the data. I sometimes hear researchers who favor one particular interpretation of the facts claim that science is on their side. But I learned from my father that science doesn't take sides. It's an impartial method for evaluating all the available data. The question is not whether science is on your side, but rather whether you are on the side of science.

So how can we test scientifically whether or not NDEs are real? On the surface, it may seem ridiculous to question whether someone's experience 'really' happened. Philosopher Abraham Kaplan tells a story about a man who goes to a distant land and returns claiming that he has seen a strange and wondrous beast: a camel. This animal is able – so he says – to travel for days across the hottest desert without water! The scholars in his hometown are astonished and puzzled. They tell him, 'We don't know whether or not such an animal can possibly exist, but we'll hold a meeting to decide whether or not, given what we know of biology, such an animal could be real.' The traveler then replies, '*Could* be real? But I tell you, I saw it!'

As psychologist Bob Van de Castle put it, if you get hit by a truck, you *know* that you were hit by a truck, and no amount of skepticism from others will ever convince you that the truck was only imaginary. I haven't yet been hit by a truck, but I was hit almost a half century ago by Holly's insistence that she had seen the stain on my tie while she was unconscious in another room. I didn't know how to make sense of that event, but I couldn't pretend it never happened, and I couldn't dismiss it as a misperception or the product of my imagination. What about all the incredible NDE stories that didn't happen to *me*, but that I've heard from other people? How could I determine whether they *really* happened?

As I noted earlier with regard to out-of-body visions, they are challenging to verify. Recall that Jan Holden reviewed ninety-three reports of out-of-body perceptions during NDEs, and found that 92 percent were verified by outside sources as completely accurate, while 6 percent contained some error, and only 1 percent was completely wrong. Obviously, there will be some claims from experiencers that may not be accurate descriptions of what really happened. But the fact that some NDE stories may be erroneous, or even made up, does not discount all NDE reports. Jalāl ad-Dīn Rumi, the thirteenth-century Sufi mystic, wrote that there wouldn't be counterfeit gold unless there was also real gold. So, too, there would not be fake NDEs unless there were real NDEs. The question is how to tell them apart. I don't know how to test the validity of stories about otherworldly realms. But I *can* test the reports of seeing things in our physical world.

One way to do so is to ask how reliable experiencers' memories are. There are several factors that made me suspect memories of NDEs might *not* be reliable. First, many NDEs occur in cardiac arrest, which often causes amnesia for events around the time the heart stops. Second, NDEs sometimes happen to people who have taken psychedelic drugs, which can interfere with memory. Third, NDEs usually occur in traumatic situations, which are known to influence the accuracy of memories. Fourth, they usually include strong positive emotions, which may influence memory. And finally, NDEs are sometimes reported long after the event, which often reduces the detail and vividness of memories. All of these factors raise questions about the reliability of memories of NDEs.

Some researchers have speculated that NDE stories are embellished over time, and specifically that memories of NDEs become more blissful as the years pass. Because I've been studying NDEs for four decades now, I've been able to address this question. Starting in 2002, I began tracking down people I had interviewed about their NDEs in the early 1980s, and asking them to describe their NDEs for me again. What I found was that the accounts had *not* become more blissful over time. In fact, there were *no* differences between what the experiencers told me in the

1980s and what they told me decades later. This suggests that experiencers' memories of their NDEs are reliable. And by extension it also suggests that studying experiences that happened years ago is as valid as studying recent NDEs.

Another important question about NDE accounts is whether they're influenced by people's beliefs. We know that cultural conditioning and expectations influence how experiencers interpret their perceptions. Remember when the trucker, Dominic, labeled his tunnel a 'tailpipe'? Do people experience in their NDEs just what they expect or hope will happen? I was able to test that idea as well. Ian Stevenson, my mentor at the University of Virginia, had been collecting what we now call NDEs for years before Raymond Moody introduced that term to Americans. Ian had filed them under various categories, such as 'out-of-body experiences,' 'deathbed visions,' and 'apparitions,' based on their most prominent features.

I selected two dozen of Ian's most complete cases from the 1960s and early 1970s, and in each one I rated the presence of the fifteen common features that Raymond described. Then with the help of Geena Athappilly, a medical student who was working with me, I selected two dozen NDE accounts I had recently collected myself, matching each one to one of Ian's cases in terms of the experiencer's age, race, gender, religion, reason for coming close to death, and how close they had been to death medically. With a single exception, every one of the NDE features Raymond described – such as leaving the body, feeling peace, meeting others, a 'being of light,' music, and a life review – were included just as often in the accounts collected *before* Raymond wrote his 1975 book, which spelled out what happens in NDEs, as they were in the recent narratives. The only exception was a sense of traveling through a tunnel, which was reported more often in the recent accounts. But recall that I had left the tunnel out of my NDE Scale because other researchers had suggested that it was something we make up after the fact, in order to explain to ourselves how we get from one situation to another, with no other awareness of how we got there.

Geena and I also rated how many accounts from Ian's collection and from my recent cases included the aftereffects Raymond had described following NDEs – things like changes in values, decreased fear of death, belief in continuation after death, and difficulty telling others about the experience. Again, all of these aftereffects were mentioned just as often in narratives collected *before* Raymond wrote his book in 1975 as they were in recent NDE accounts. So NDE accounts have not changed over decades, and they do not appear simply to reflect familiar models of what happens at death.

But are these consistent accounts memories of events that really happened, or are they memories of events that were only imagined? In fact, most experiencers are quite certain about the reality of their NDEs and describe them as 'realer than real' or 'more real than anything else I've ever experienced.' A survey of more than six hundred experiencers by radiation oncologist Jeffrey Long found that 96 percent rated their NDEs as 'definitely real,' while none rated them as 'definitely unreal.' The participants in my research echo this conviction in the absolute reality of NDEs. Among all the experiencers I've studied, 71 percent said their memories of their NDEs are clearer and more vivid than their memories of other events, while only 3 percent said they were less clear or less vivid.

Jayne Smith, who had an NDE at age twenty-three during a bad reaction to anesthesia during childbirth, told me, 'Never, ever did I think it might have been a dream. I knew that it was true and real, more real than any other thing I've ever known.' LeaAnn Carroll developed a massive blood clot in her lung at age thirty-one that stopped her heart. She said about her NDE, 'My death experience is more real to me than life.' Nancy Evans Bush, who had an NDE at age twenty-seven during a bad reaction to nitrous oxide, said, 'Yes, it was more real than real: absolute reality.' Susan Litton, who had an NDE at age twenty-nine, told me, 'There was no sense of doubt whatsoever. Everything had a sense of being "more real" than anything that would normally be experienced in the physical world as we know it.' Chris Matt, who had an NDE when he rolled his car over at age twenty-one, said, 'I have no doubt that it was real. It was *vastly more real* than anything we experience here.' Yolaine Stout, who attempted suicide at the age of thirty-one, said, 'This was more real than anything on earth. By comparison, my life in my body had been a dream.'

There are ways to distinguish whether memories are of real events or of fantasies. Lauren Moore, a psychiatric trainee, and I used the Memory Characteristics Questionnaire (MCQ), a widely used scale that was designed to differentiate memories of real events from memories of fantasies or dreams. The MCQ looks at aspects of memories that are reliably different in memories of real or imagined events, such as the clarity and detail of the memory, its sensory aspects, memory for the context of the event, processing of thoughts when recalling the memory, and the intensity of feelings associated with it. I asked people who had come close to death to rate their memories of three different experiences. The first was their close brush with death. The second was another real event that occurred around the same time. And the third was an imagined event, also from around the same time.

What Lauren and I found was that, for people who had NDEs, their memories of the experience were like memories of real events, but not like memories of imagined events. In fact, NDEs were remembered as *more real* than real events, just as real events were remembered as more real than imagined events. Their memories of the NDE had more detail, more clarity, more context, and more intense feelings than memories of real events. And that is exactly what people had been telling me for decades – that their NDEs were more real to them than everyday experiences. On the other hand, for people who had come close to death but *didn't* have an NDE, their memories of the event were *not* recalled as more real than other real events. Two other research teams, in Belgium and in Italy, came up with the same results. In addition, the Italian team also measured brain waves while experiencers were remembering their NDEs. They found that the experiencers' brain waves were like those associated with recalling real events rather than recalling imagined events.

So it appears that scientific investigation of memories of NDEs confirms that they are consistent over time, that they are not dependent on familiar models of what's supposed to happen at death, and that they look like memories of events that really happened. But if NDEs are real experiences and not hallucinations or fantasies, how do we explain them? This question led me next into a search for what happens to the brain during a close brush with death.

The Biology of Dying

How do scientists explain near-death experiences? I'd learned a lot about the brain and how its different parts work, and I wondered whether there might be *specific areas of the brain* that are linked to NDEs. Several researchers have tried to locate regions that might cause an NDE when they're stimulated with an electrical probe. They've most often targeted the temporal lobes, on the sides of the head, just below the temples. Some researchers have argued for NDEs being associated with the right temporal lobe, while others have argued for the left.

Many scientists have cited the pioneering work of neurosurgeon Wilder Penfield as evidence that abnormal electrical activity in the temporal lobe can cause out-of-body experiences. In the 1950s, Penfield worked at McGill

University's Montreal Neurological Institute with patients who had seizures – sudden electrical discharges in the brain – that didn't respond to medications. For many of these patients, the only treatment available was surgery to remove the part of the brain causing the seizure. Penfield's technique for finding the exact spot in the brain where the seizures originated involved opening the patients' skulls and touching different parts of their brains with a mild electric current. Surprisingly, this doesn't hurt at all, because there are no pain receptors in the brain. Because of this, Penfield was able to probe patients' brains while they were fully awake, so they could report to him what they were experiencing.

With this technique, Penfield not only found where the seizures originated but also identified what each part of the brain did when it was stimulated. He was the first person to map out the brain, indicating which areas controlled movements of the patient's fingers, lips, and so on. He also identified discrete brain areas that gave rise to various sensations when they were stimulated, like feelings of heat or cold, or a certain smell, or hearing a sound like a certain song, or seeing scenes from the past as though the patients were watching a movie.

Penfield is widely believed to have produced out-of-body experiences and other NDE-like phenomena by electrically stimulating various points in the temporal lobes of his conscious patients. In fact, though, only two out of the 1,132 patients he stimulated reported anything that was even *vaguely* like an out-of-body experience. When

Penfield touched an area on the right temporal lobe of a thirty-three-year-old man having an operation under local anesthesia, the man suddenly said, 'That bittersweet taste on my tongue.' The patient was confused and made tasting and swallowing movements. Penfield then turned the electrical current *off*, and the man said, 'Oh, God! I am leaving my body.' He looked terrified and made gestures for help. Penfield then stimulated deeper in the temporal lobe, and the man said he was spinning around and felt as though he were standing up.

In the second case, when Penfield touched a woman's temporal lobe with his electric rod, she said, 'I have a queer sensation as if I am not here.' As he continued the electrical stimulation, she added, 'As though I were half and half here.' He then touched another area of the temporal lobe, and she said, 'I feel queer,' adding that she felt as though she were floating away. As Penfield continued the stimulation, she asked, 'Am I here?' He then touched a third area in the temporal lobe, and she said, 'I feel like I am going away again.' She had a sense of unreality, not at all like the experiencers' sense of NDEs being 'more real than real.' She felt as though she were somewhere else and yet still in her environment. Neither of these patients who felt they might be leaving their bodies reported seeing themselves from above, as often happens in NDEs.

Despite these suggestive but hardly convincing anecdotes from Penfield's work, later studies of patients with seizures have found that a sense of leaving the body can be reported by patients with a variety of seizure types affecting different parts of the brain. All in all, there seemed to be scant evidence for the temporal lobe being involved in a sense of leaving the body.

Other neuroscientists have argued that NDEs and similar experiences are associated with different parts of the brain, including the frontal lobe, the parietal lobe, the thalamus, the hypothalamus, the amygdala, and the hippocampus. Neuroscientist Mario Beauregard and his colleagues at the University of Montreal measured brain activity in people who had had NDEs. They scanned these experiencers' brains while they were trying to relive their NDEs during meditation. Beauregard found that there was no one part of the brain that was associated with NDE memories. Rather, several different parts of the brain became active when the NDEs were remembered.

These conflicting findings of NDEs being associated with different parts of the brain – or with several parts working together – left us without a definitive answer. Since other researchers had suggested that people with seizures sometimes experience the sensation of leaving their bodies, I decided to look more closely at out-of-body phenomena in this particular population. My hope was to see whether electrical discharges in specific parts of the brain would be associated with NDE-like sensations more than in other parts of the brain. I approached Nathan Fountain, the neurologist who ran the epilepsy clinic at my hospital, to get his permission to interview his clinic patients about their experiences during their seizures. He sat at the desk in his office, which overflowed with patient charts and other papers, as I explained to him what I had in mind. Though he listened attentively and was perfectly cordial, I got the feeling my proposal did not excite him.

'I don't think you'll find what you're looking for,' he said, shaking his head. 'Seizures disrupt the normal functioning of the brain, so the patients' consciousness is impaired and they won't be able to tell you anything. If they're having a generalized seizure, they're unconscious the whole time, so they can't be having an "experience" of any kind. In fact, many of them don't even know that they had a seizure.' He shrugged with a smile, and added, 'And if they're having temporal lobe seizures, even if they did have some kind of experience, they wouldn't be able to tell you about it. Because their hippocampus is seizing and not able to function normally, they wouldn't be able to form a memory of it.'

'I've read accounts of people having a sensation of leaving the body when the temporal lobe is stimulated,' I argued. 'Two of Wilder Penfield's patients reported something like that when he stimulated their temporal lobes with an electric current.'

'But Penfield's patients were awake,' Nathan protested, 'and he used a very mild current. A seizure is completely different. Most people are unconscious during their seizures and have no memory of it after it's over. They're not going to be able to tell you anything.'

'You may be right,' I conceded, 'but would you be willing to let me talk with some of your clinic patients and ask them?'

He hesitated, then asked, 'How would I explain to them that I want them to talk to a psychiatrist?'

'Just say that you know having uncontrolled seizures can be stressful, and that one of your colleagues is doing a study of what people feel during a seizure and how they deal with it.'

I could see him weighing the idea in his mind. 'Some of them may welcome the chance to talk with someone about what it's like to have seizures.'

I thought his resistance was weakening, so I pressed on. 'Why don't we interview two or three patients together, so you can hear for yourself what they say and how they feel about talking with me?'

'Okay,' he said. 'If you come to the clinic next Monday afternoon, I'll sit in with you as you interview some patients. I'll schedule fewer patients, so we can spend more time with each one and discuss their seizures in detail. Then we'll see where it goes from there.'

The first couple of patients we interviewed had no memory at all of their seizures – just as Nathan expected. They just blacked out and then awoke some time later, feeling confused or exhausted. The third patient, Marie, was a young secretary who'd been having poorly controlled seizures since she'd had meningitis in college. She was not at all reluctant to talk with me. I asked her about the last thing she usually remembered before her seizures started.

'Ew,' she began, 'I usually get this strong odor, sort of like dirty gym socks. It lasts just a few seconds, but I know it means I'm going into a seizure.'

'And what's the next thing you usually remember when the seizure is over?' I asked.

'I'm groggy for a while afterwards,' she said. 'I usually

just lie there, on the floor or wherever, trying to remember where I am and what I was doing. I don't know how long that lasts, but I usually wake up pretty slowly, and then I'm still tired for a long while.'

'And between those two times,' I asked, 'after the smell of gym socks and before you wake up feeling groggy, what do you remember of that period?'

'You mean during the seizure itself?' she asked.

I nodded. 'Mm-hmm.'

She shook her head. 'I'm usually out cold. Everything's a blank.' She looked uncertainly at me, then at Nathan, and then back at me. 'But there was one time when I thought I remembered something.'

'One time?' I repeated, nodding.

'Yeah,' she continued hesitantly. 'It seemed like I was watching my body shaking on the floor.' She looked back at the neurologist and went on. 'Of course, I know that couldn't be, but after it was over, I had this memory of looking down on my body, and my arms and legs were jerking.' She stopped, as if embarrassed to go on.

'And that happened just once?' I asked.

She nodded. 'Yeah, it was years ago, before I was married. I was alone in my apartment, reading a magazine, when it happened. It was confusing, but I just thought it must have been my imagination.'

I looked over at Nathan, who was smiling. That was typical of him, to react to an unexpected event with amusement rather than shock.

'And how did that experience affect you?' I asked Marie.

She shrugged. 'It didn't, really. I never forgot it, but it

didn't affect me in any way.' She paused, looked first at Nathan and then at me, and added, 'No one's ever asked me before what it's like for me during a seizure. I'm glad to be able to talk about it.'

'Is there anything else you can remember about that experience, or anything else you want to tell us, or ask us?'

'No, that was it,' she said. 'It was just a strange thing that I've never forgotten.'

After I thanked her for speaking with us and ended the interview, I turned to Nathan.

'Okay,' he said, 'we don't need to interview anyone else. Let's do the study.' As with most doctors, what he saw and heard was more persuasive than what he'd been taught he *should* be seeing and hearing.

My research colleague Lori Derr and I then interviewed a hundred patients in the epilepsy clinic about their experiences during their seizures. We started by asking them what they could recall just before and after the seizure, and then during the seizure itself. After they responded, either with their memories of the seizure or with no memory, we asked about all the items on the NDE Scale, on the chance that some of them may have experienced some NDE features but not considered them worth mentioning. For example, we would ask, 'During your seizures, have you ever felt your sense of time change?' Many of them had no memories of any experiences at all during their seizures. But slightly more than half of the patients did remember *some* fleeting sensation, like a strange smell or sound. And as Nathan was quick to point out, those memories could have been of sensations they'd had just *before* the seizure started.

Out of the hundred patients, seven reported they lost their sense of time, one reported a feeling of peace, and one reported seeing a bright light. And out of the hundred patients, seven reported having had an experience that was at least vaguely like leaving the body during a seizure. Almost all of them reported just one such experience during decades of seizures, and couldn't remember exactly when it was ('about fifteen or twenty years ago, I think'), making it impossible to identify specific medical details that might have been associated with the experience. In addition, they almost all told us that they knew these experiences weren't real.

For example, Maryann, a thirty-year-old woman who reported 'two or three out-of-body experiences' during seizures, added, 'I know it couldn't be real. It's too far-fetched. I didn't actually see my body or anything else. It may be my imagination.' And Mark, a forty-two-year-old man, told me, 'I don't remember any of the details of it. It's like a dream. Obviously, I don't think that sort of thing can really happen.' These patients usually gave only the vaguest descriptions of losing awareness of their bodies. With one exception, they didn't report seeing their bodies from the outside.

In contrast to these patients with seizures, people who report NDEs almost always insist their experiences were real. They adamantly deny the possibility that their memories could be the result of their imaginations or a dream. They often look down on their unconscious bodies, and they remember the whole experience in sharp detail.

Only one of our hundred patients reported having left her body often during seizures. But although she described being able to see her body quite clearly from a position above it, her account of these experiences was very different from what most people describe in NDEs. Kirsten was a twenty-eight-year-old graduate student in psychology who'd been having seizures from birth as a result of an inborn brain malformation near the top of her head. The abnormality was in the midline between the right and left parietal lobes, and caused her to have seizures a few times a month for her entire life. During her seizures she appeared to be unconscious for a short time, usually less than a minute. To other people, she seemed to get a blank look and just stare, not responding to what was happening around her. If she happened to be walking, she would continue to walk in a straight line as if she was unaware of where she was going. To Kirsten, though, it seemed as if she knew what was happening during her seizures. I asked her to describe what they were like for her.

'If I'm reading,' she began, 'the letters on the page become nonsense characters. And if I'm talking with a friend, what comes out of my friend's mouth no longer sounds like words. I can think of things *I* want to say, but I can't make sounds except for gibberish.' She shrugged her shoulders and shook her head.

'And then what?' I asked.

'And then I'm several feet above my body, looking down.'

That took me by surprise. I had already interviewed dozens of patients with seizures and had not yet heard a single clear account of leaving the body. But Kirsten described it as a routine part of every seizure. 'Tell me about that,' I said. 'What's that like for you?'

'It's terrifying!' she said.

'Terrifying?' I repeated. Again, I was surprised, because most experiencers describe feeling relief or a sense of freedom when they leave their bodies in NDEs. But as I later learned, Kirsten's response was not unusual. Other researchers have reported that patients who described a sense of leaving their bodies during seizures commonly report intense horror or fear.

'Yeah,' she went on, 'I'm afraid that something will happen to my body while I'm not in it.'

'Like what?' I asked.

'I don't know.' She shook her head. 'I'm just afraid something will happen if I'm not there to protect it.'

'Has anything ever happened?' I asked. 'Have you ever gotten hurt while you were out of your body?'

'Not really hurt,' she said, slowly. 'A few weeks ago, I was pushing a shopping cart down the aisle of a supermarket when I had a seizure. While I was out of my body, it ran the cart into a display rack and knocked over a tower of cans.'

'What happened?'

'Well, that snapped me back into my body, and I picked up the cans and restacked them. People stared at me, and finally one person came over to help.'

'So what snapped you back into your body? Was it hitting the rack, or the sound of the cans crashing?'

'As soon as I felt the cart hit the rack, I snapped back into my body. That's what usually happens.'

'That's what usually brings you back from a seizure?' I asked.

Kirsten nodded. 'Anything that touches or jars my body can do it. If I'm talking to one of my girlfriends and I suddenly stop in midsentence, she'll usually call my name and tap my arm to bring me back.' She smiled and added, 'Of course, none of my boyfriends have ever even noticed when I've had a seizure.'

Her description of being pulled back into her body by being touched did ring a bell. I remembered the NDE described by Sir Alexander Ogston, the Scottish surgeon hospitalized with typhoid fever, who was also repeatedly pulled back into his body when it was jarred: 'I was conscious that my mental self used regularly to leave the body . . . until something produced a consciousness that the chilly mass, which I then recalled was my body, was being stirred as it lay by the door. I was then drawn rapidly back into it, joined it with disgust, and it became *I*. When it was again left I seemed to wander off as before . . . until something again disturbed the body where it lay, when I was drawn back to it afresh, and entered it with ever-growing repulsion.'

'And what happens if no one or nothing touches your body when you're out?' I asked Kirsten.

'Oh, I'm never gone long. I can't do anything myself to get back into my body, but in a few seconds, or sometimes maybe a minute or so, the seizure runs its course and I pop back in.' She paused, and then added, 'And then I'm fine. It's as if nothing happened.'

'Do you ever experience anything else when you're outside your body?' I asked.

'Like what?' She wrinkled her forehead.

'Like, do you ever see or hear anything?'

She shook her head. 'No, I just try to focus on my body in case anything happens to it.' She gave me half a smile. 'Like ramming into a display of cans.'

'And what else do you *feel* when you're out of your body? You said it can be terrifying, but do you ever have pleasant feelings, or feel like you're going someplace else?'

'Pleasant feelings? Are you kidding?' Her eyes widened. 'It's scary. I don't like it at all. I never have. I can't wait to get back into my body and act normal again.'

Kirsten's protective attitude toward her body was very different from most experiencers, who usually seem unconcerned about their physical bodies during their NDEs. I recalled Jill Bolte Taylor's description of feeling herself outside her body during her stroke: 'I felt like a genie liberated from its bottle. The energy of my spirit seemed to flow like a great whale gliding through a sea of silent euphoria. Finer than the finest of pleasures we can experience as physical beings, this absence of physical boundary was one of glorious bliss.'

Turning my attention back to Kirsten, I nodded and asked her, slowly, 'Kirsten, you're a psychology grad student. You know a lot about people and our brains. How do you understand what happens when you leave your body? How do you make sense of that?' I was asking partly to see how she understood it, and partly hoping she would help me understand it.

She just shrugged and shook her head. 'It's just something that happens when I have a seizure. I've never tried to figure out *how* it happens. I just want it to *stop* happening.'

I pressed a little more: 'And what is it that leaves your body during the seizures?'

'Me!' She shrugged again, clearly not wanting to think about this, and getting impatient with me.

I gave it one more try: 'Do you think that you *really* go out of your body, or is it just a trick of your brain when it's having a seizure? I mean, if you were to have a seizure right now and rise up above us, would you be able to see the bald spot on the back of my head?'

She laughed. 'Of course!' she said. 'Are we finished? I don't want to be late for my class.'

I was struck by Kirsten's apparent lack of curiosity about her out-of-body experiences. The sensation of leaving the body behind raised all sorts of questions in me, but it didn't seem to faze her a bit. But then, Kirsten's childhood had been very different from mine. I grew up strongly identifying with my body. When my body felt the adrenaline rush of a footrace, or when it felt tired after a long day, or when it felt hungry between meals, or when it felt that tingly warmth of a hot bath, it seemed that I *was* my body. But Kirsten had repeatedly experienced the sensation of leaving her body behind since she was an infant, and it just seemed like a natural state for her. Being able to separate from her physical body was not a mystery for her, but simply a routine part of her life experience. And that was a very different attitude from what I'd heard from experiencers, who were generally startled to find themselves out of their bodies for the first time during their NDEs.

In keeping with Beauregard's findings of brain activity in experiencers recalling their NDEs, we found that seizure experiences that sounded even vaguely like NDEs or out-of-body experiences were not associated with any particular lobe of the brain, nor with either the right or left side. Kirsten, the only patient in our study who recalled what sounded like a clear out-of-body experience, had a defect in the midline of her brain. And although she claimed she really left her body and saw things accurately from another location, she reported no other features of a near-death experience and nothing that sounded or felt at all pleasant or uplifting.

Despite the common belief among some scientists that unusual electrical activity in the temporal lobe, like that caused by epileptic seizures or stimulation, can provoke experiences like NDEs or out-of-body experiences, we didn't find that to be true. Other researchers have reported that stimulating the temporal lobe of a patient with an electric current can produce a feeling of the body being distorted, or even a sensation of leaving the body. But there are many important differences between these sensations induced by electrical stimulation and the out-of-body experiences associated with NDEs. Perhaps the most crucial is that patients whose brains are being stimulated describe these sensations as unrealistic dreamlike events, not as things that are really happening, whereas people describe their NDEs as undeniably real events. It's a bit like watching a war movie compared with actually fighting in a battle. Someone describing being in battle and someone describing a war movie may report seeing similar images and perhaps feeling similar emotions, but fighting in a battle is experienced as obviously real and watching a movie is experienced as an imitation of the real thing.

Having abandoned the idea that NDEs were associated with a particular *part* of the brain, I next considered the possibility that they might be related to electrical activity of the brain *as a whole*. Is it possible that at or near the point of death, there's still enough electrical activity in the brain to produce a vivid and elaborate experience? The medical literature did not encourage that idea. Decades of clinical experience and research have established that brain activity decreases within six to seven seconds of the heart stopping. And after ten to twenty seconds, the electroencephalogram (EEG) goes flat, indicating no activity in the cerebral cortex – the part of the brain responsible for thoughts, perceptions, memory, and language. Analysis of the EEGs of people after life support is withdrawn show that the brain's electrical activity in such cases actually stops *before* the heartbeat stops and before blood pressure ends – and after the heart stops there is *no* well-defined EEG activity. That seemed to answer my question about whether NDEs could be related to electrical activity in the brain.

I also wondered whether NDEs might be elaborate fantasies or dreams that we create in a crisis to distract ourselves from the pain or terror of the close brush with death. Neurologist Kevin Nelson put this idea in physiological

terms by suggesting that the kind of brain activity associated with dreams, commonly called rapid eye movement
(REM) brain activity, can intrude into our waking thoughts
in a near-death crisis, producing dreamlike thoughts and
images. He did a study that he thought showed a high rate
of REM intrusion symptoms in experiencers. However, it
turned out that the rate found in this study was no higher
than the rate of the same REM intrusion symptoms in a
random sample of the general public who didn't have
NDEs. Another problem with this explanation is that
many NDEs occur in people under general anesthesia,
which suppresses REM brain activity. Furthermore, measurements of REM brain activity in people who've had
NDEs show that it's actually *lower* than in other people.
Finally, an Italian research team found that experiencers
remembering their NDEs did not have brain wave patterns
typical of recalling fantasies or dreams, but had brain wave
patterns typical of memories of real events. So it seemed
that near-death experiences are decidedly not like any
dream.

What about chemical changes in the brain? Might they be
involved in NDEs?

In terms of brain chemistry, I wondered whether
decreased oxygen in the brain might be a factor in causing
NDEs. People report similar experiences no matter how
they came close to death, so it made sense to look at the
bodily events that occur in all near-death situations. And
no matter how we come close to death, one of the final

events is that our heartbeat and breathing stop, cutting off the flow of oxygen to the brain. But the medical literature has established that decreased oxygen is a very unpleasant experience, particularly for those people who report perceptual distortions and hallucinations. The fear, agitation, and combativeness typical in people with decreased oxygen are very different from NDEs, which are usually peaceful, positive experiences. But the most conclusive evidence for me came from research that actually measured oxygen levels in people during medical crises. That research consistently showed that NDEs are associated either with *increased* oxygen levels, or with levels the same as those of non-experiencers. No study has ever shown *decreased* levels of oxygen during NDEs.

My next thought was that NDEs might be related to the medications given to people in medical crises. Certainly most patients in medical crises are given a variety of medications. But again, the medical literature did not provide any support for this idea, either. In fact, research showed that patients who are given medications in fact report *fewer* NDEs than do patients who don't get any medication. One reason I found this model intriguing is that NDEs are similar to some unusual experiences associated with psychedelic drug use. Researchers have found some common descriptions in reports of NDEs and reports of hallucinogenic trips, most often those associated with the anesthetic ketamine and with DMT (dimethyltryptamine), a chemical that is found in nature and can produce euphoria and visions.

I was recently part of a multinational research team that analyzed language usage and language structure in 625 accounts of NDEs and compared them to almost fifteen thousand accounts of unusual experiences of people taking any of 165 different drugs. We found that the drug states most similar to NDEs were those associated with ketamine. However, we were careful to note that other common effects of ketamine *don't* appear in NDEs, which suggests that NDEs are not simply an effect of the drug. In a similar vein, Karl Jansen, the neuroscientist who has most fiercely promoted the ketamine model for NDEs, concluded after twelve years of research that he viewed ketamine as 'just another door' to NDEs, and not as actually producing them.

If NDEs are not associated with medications given to people, might they be related to chemicals *produced* by people in crisis? We know that our brains produce or release a number of chemicals to help the body cope under stress. The chemicals I thought might be most likely to be associated with NDEs were endorphins, the 'feel-good hormones' that produce the 'runner's high' in marathon runners, and that are known to reduce pain and stress. Other scientists have suggested that NDEs might be connected to serotonin, adrenaline, vasopressin, and glutamate, all of which are chemicals that transmit signals between nerve cells. But in spite of the theoretical reasons for thinking that brain chemicals might be involved in NDEs, at this point, there has been no research looking into this possibility. And I don't expect any such research to be done in the near future.

Bursts of these chemicals in the brain tend to be very short-lived and localized, so in order to find them, we'd have to look at exactly the right time at exactly the right place in the brain – and as I discovered, we don't even know where in the brain to look.

None of these brain-based models that seemed at first glance to be good candidates for explaining NDEs turned out to be adequate. Exploring these plausible stabs at explanations reminded me of the ancient Indian parable of the blind men and the elephant, which dates back at least to the Buddhist Udana about 2,500 years ago. In this story, a group of blind men who have never before come across an elephant try to grasp its essence by feeling it. One grabs the trunk and says that an elephant is like a water hose. Another grasps a tusk and says an elephant is like a spear. A third feels a leg and says an elephant is like a pillar. Still another touches an ear and says an elephant is like a fan. Each one comes away with a reasonable analogy for what an elephant is like based on his limited subjective perception. But none of them understands the entire elephant.

In some ways, our inadequate models to explain NDEs are also reasonable analogies based on limited subjective perceptions of one or another feature of NDEs. For example, the blissful emotions in NDEs are a bit like the good feelings produced by endorphins, and the visions in NDEs are in some ways like the hallucinations produced by ketamine, and the life review in NDEs could conceivably be compared to the fragmentary memories that can be provoked by temporal lobe stimulation. But while each of these models may give a rough analogy for one limited feature of NDEs, none of them adequately describes the entire experience.

Nevertheless, as a doctor, I knew that I wouldn't be able to understand the mystery of NDEs unless I learned more about what was going on in people's brains at the time they were having near-death experiences. That proved to be challenging. Most NDEs happen far from any medical care, where there's no chance for any medical monitoring, let alone brain imaging. Even those NDEs that occur under close medical supervision generally happen in critical situations, when hospital staff are thinking only of resuscitating the patient's heart, and not of scanning the brain. But every once in a while, we do manage to get a rare glimpse into the brain during a near-death experience.

10

....

The Brain at Death

The fifty-four-year-old neurosurgeon awoke suddenly at 4:30 in the morning with a severe headache and back pain. Within four hours, he was unconscious and his family couldn't wake him. His wife called emergency medical services when he started seizing, and he was rushed to the emergency room of the local hospital. His neurological examination, which included a brain scan, showed widespread damage to his cerebral cortex – the part of the brain associated with thinking, perceiving, forming memories, and understanding language. It also showed damage to his brain stem – the part that regulates breathing, swallowing, heart rate, blood pressure, and whether you're awake or asleep. The emergency room physicians judged him to be near death, with little chance of surviving. He spent the next several days in a deep coma, with antibiotics, medications to prevent further seizures, and a ventilator to help him breathe. On his third day in the hospital, a repeat brain scan showed pus filling much of the area inside his skull. He didn't respond at all when people spoke to him, pinched him, or stuck him with needles.

He turned out to have a rare acute bacterial infection of his brain, one associated with a 90 percent fatality rate. But on day six of his coma, he surprised everyone by opening his eyes. He was awake but confused. He had difficulty controlling his arms and legs, and he didn't understand where he was or recognize his family. He had no memory of his life before the coma, and couldn't form words. But he awoke with vivid memories of an elaborate experience in a very different kind of environment.

To his doctors' surprise, his memories and ability to speak began to come back, and within a few more days he was able to describe what he recalled of his experience in the coma. He reported being lifted by a slowly spinning, clear white light associated with a musical melody, into a rich and ultra-real environment with light and colors beyond the normal spectrum, in which he seemed to be aware but without a body. He described flowers, waterfalls, beings dancing in joy, angelic singing, golden orbs swooping across the sky, and a young woman who seemed to be some kind of guide, though she never spoke in words. He then rose from that realm into a 'higher' one, of infinite inky blackness overflowing with the healing power of an all-loving deity, for whom 'the term "God" seemed too puny a word.'

He also reported having seen specific people who were

not family members praying around his hospital bed, which would normally be prohibited for a patient in the intensive care unit. His family and the hospital staff confirmed that those people he mentioned had indeed been present, on a day when his score on a standard coma scale showed severe brain impairment. He also described to his family a detailed physical description of the woman who had accompanied him in the 'ultra-real' environment. Four months later, the patient, who had been adopted at birth, met his birth family, and they showed him a photograph of a sister he had never met, but who had died ten years earlier. He was stunned to recognize her as his guide in his NDE.

Some of you may recognize this story as the experience of neurosurgeon Eben Alexander, who became one of the most widely known experiencers. Following his son's advice, he refrained from reading anything about NDEs to avoid biasing his own memories, until he had written down everything he could remember about his own experience – a process that took two years.

On the second anniversary of his NDE, Eben drove to the University of Virginia, accompanied by his son, who was at that time a college student majoring in neuroscience. Having written an impressive twenty-thousand-word account of his NDE, he now wanted to start exploring what it meant. Everything he had known as an academic neurosurgeon told him that he could not have had – or formed a memory of – any kind of experience during his coma, let alone a vivid and 'ultra-real' one. And yet he did.

When he approached me, Eben was not so much distressed by this puzzle, as I would have expected, but rather energized by it, eager to try to make sense of it. He drove the hour and a half to the University of Virginia to speak with me about how to start exploring the field of near-death research. I asked his son what he, as a neuroscience major, made of his father's experience. The young man just shook his head. After a pause, he said, 'I don't know, but this is not the father I grew up with.'

Some popular writers have claimed that Eben was never really in danger of dying, that his apparent coma was really just sedation due to the medications he had been given. Having been trained to be skeptical, I was unwilling to accept at face value either Eben's own account of his medical condition or the dismissive speculations of his critics. For that reason, I obtained from the hospital the complete medical record of his hospital course. In addition to reviewing the record myself, I had two other physicians, Surbhi Khanna and Lauren Moore, independently review the more than six hundred pages to evaluate Eben's medical condition. We had planned, after all three of us had completed our independent assessments, to meet as a group to try to resolve any differences in our conclusions. But when we came back together to compare notes, we found there were no differences to resolve, because the medical record was very clear and left no doubts.

Eben's CT scans showed his brain cavity swollen with pus, and notes from his doctors documented their expectation that he would never wake up – or if he did, would never again be able to speak or function on his own. All three of us concluded independently that he had been extremely close to death, with a brain as disabled as it could have been, and that while that was happening, he witnessed things that a comatose person should not have been able to perceive. The data showed that his coma had not been related to the drugs he was given. The medical record noted that he was rapidly falling into a coma by the time he reached the hospital, before he received any medication. And six days later, he came out of the coma before the medications were stopped.

According to our current understanding of how the brain works, it should not have been possible for Eben to have had any experience at all during his deep coma – let alone the most vivid and memorable experience of his life. And yet, he did. And furthermore, he is not the only person to have had a vivid and profoundly memorable experience during such a medical crisis.

So how do we make sense of these experiences that seem to contradict what we know about brain function? At this point, we need to step back and talk about the difference between the *mind* and the *brain*. Right now, you are reading this page and thinking about what you're reading – and maybe you're also feeling an itch on the bottom of your foot. Those feelings and those thoughts are examples of what we mean by consciousness: being aware of yourself and the world around you. Your own consciousness is both the most complex puzzle for humans and also the simplest, most self-evident fact. Nothing is more obvious and undeniable than the fact that you are conscious – that you're aware of what you're doing and what's going on around you.

Your mind is the sum total of all your conscious thoughts, feelings, desires, memories, hopes, and so on. Your brain, on the other hand, is that mass of pink-gray matter inside your skull, made up of nerve cells or neurons, and supporting cells or glia. We know that our minds and our brains are connected, but after thousands of years of personal observations and hundreds of years of research, we still don't know what exactly that connection is.

Over the centuries, a wide variety of models have been proposed for the relationship between the mind and the brain. Put in the simplest terms, the most common models assume either that the mind is a product of the brain, or that the mind and the brain are two separate things. Neither of these models can explain the mind-body relationship completely. On the one hand, if the brain produces the mind, we have no idea how it could do that. On the other hand, if the mind is not produced by the brain, where does it come from? And how do we explain the close link between mind and brain? Philosophers and scientists have been debating this issue for centuries. The reason we still argue about the mind-brain question is that we still haven't come up with an answer that seems to work – at least not one that works all the time.

Most of us assume that the workings of the mind can be explained by the physical brain. That is, 'The mind is what the brain does.' In other words, our consciousness, perception, thinking, memory, emotions, and intentions are produced by electrical and chemical changes in the brain. We have a lot of evidence for that viewpoint.

First, there is the everyday association between brain activity and mental experiences. For example, when you get drunk, or when you get hit on the head, you don't think as clearly as usual. And certainly brain diseases like strokes and seizures and concussions can affect our ability to think and feel and remember. There are also scientific experiments showing that specific mental functions are associated with activity in specific parts of the brain. For example, seeing is associated with activity in the occipital lobe, which is in the back of the brain and receives information from the eyes. And we have learned that removing certain parts of the brain interferes with certain mental experiences – so if your occipital lobe were to be surgically removed, you wouldn't be able to see, even if your eyes were still functioning normally. We have also discovered that stimulating certain parts of your brain with electrical current can cause you to have certain mental experiences. So electrical stimulation of your occipital lobe might cause you to have an experience of sight. Putting all this evidence together, it seems reasonable to think that the mind or consciousness is created by the brain.

But that's not the only way to understand the link between the brain and the mind. We have to be careful not to confuse what might simply be an association with cause and effect. The color of the sock on my left foot is usually the same color as the sock on my right foot. If you know the color of one, you can usually guess the color of the other. But the color of my left sock doesn't *cause* my right sock to be a certain color. If I happen to put a blue sock on my left foot and a brown sock on my right foot, one sock can't turn the other sock the same color.

In the same way, associations between brain activity and mental function do not necessarily mean that the electrical activity in the brain caused the thought or feeling. Maybe the thought caused the electrical activity in the brain. For example, as you read the words on this page, nerve cells in your eyes send electrical signals to the vision center of the occipital lobe of your brain and to the language center of your temporal lobe. But that doesn't necessarily mean that the electrical activity in your nerve cells is causing you to read the words on this page. Maybe your reading these words causes the electrical activity in your nerve cells.

Wilder Penfield devoted decades of pioneering research to mapping out the functions of the various parts of the brain by stimulating them with electric currents. But when Penfield stimulated the part of the brain that made his patients' arms and legs move, they didn't think that *they* were moving their limbs. Instead, they reported that they felt as if *Penfield* was forcing their limbs to move – against their will. Penfield summed this up at the end of his career: 'When I have caused a conscious patient to move his hand by applying an electrode to the motor cortex of one hemisphere, I have often asked him about it. Invariably his response was: "I didn't do that. You did." When I caused him to vocalize, he said: "I didn't make that sound. You pulled it out of me . . ." There is no place in the cerebral cortex where electrical stimulation will cause a patient to believe or to decide.'

Clearly Penfield's patients could tell the difference between their own minds wanting their limbs to move and their brains making their limbs move because Penfield was applying electrical current. They were convinced that their brains and their minds were different things.

Saying that 'the mind is what the brain does' is not quite the same as saying that 'digestion is what your stomach does.' We know how the stomach produces digestion, from the muscles in the stomach lining churning and mashing your food into small pieces to the acid and other stomach chemicals breaking the food down into nutrients we can use.

When it comes to the brain, though, we know a lot less. We can say, for example, that coordinating the movement of our bodies is part of what the brain does. We know how the brain does that by sending electrical impulses through motor nerve cells down our spinal cord to our muscles, where those impulses stimulate muscle cells to contract and move our arms and legs. But as for how the physical brain could produce thoughts and feelings and memories and conscious awareness of the world around us – we have no idea. Scientists and philosophers agree that, as philosophy professor Alva Noë put it, 'After decades of concerted effort on the part of neuroscientists, psychologists, and philosophers, only one proposition about how the brain makes us conscious – how it gives rise to sensation, feeling, subjectivity – has emerged unchallenged: we don't have a clue.'

Physicist Nick Herbert put the problem this way: 'Science's biggest mystery is the nature of consciousness. It is not that we possess bad or imperfect theories of human awareness; we simply have no such theories at all. About all we know about consciousness is that it has something to do with the head, rather than the foot.'

We don't understand how there can be mental activity – thoughts, feelings, and memories – while the brain is impaired or completely shut down, as in Al Sullivan's ability to see his surgeon flapping his arms and Bill Hernlund's ability to see his colleagues drag his body away. But we also lack basic explanations for the everyday thoughts, feelings, and memories that arise while our brains are working quite well. The dirty secret of neuroscience is that we have no idea how a physical event like electrical current or a chemical change in a nerve cell can produce consciousness.

Saying that 'the mind is what the brain does' is like saying 'making music is what a musical instrument does.' Musical instruments *do* produce musical sounds – but not by themselves. It takes something outside the instrument – a musician – to decide what sound to make and to make the instrument produce that sound. To quote Alva Noë again: 'Instruments don't make music or produce sounds. They enable people to make music or generate sounds . . . The idea that consciousness is a phenomenon of the brain, the way digestion is a phenomenon of the stomach – is as fantastic as the idea of a self-playing orchestra.'

An alternative interpretation is that the mind is not *produced* by the brain, but normally *works together* with the brain. William James, the father of American psychology, wrote more than a hundred years ago that the mind being a function of the brain can be interpreted in two very different ways. On the one hand, it can mean that the brain *produces* thought, the way a teakettle produces steam or a waterfall produces power. If that were the case with the brain and thought, then when the brain dies, it can no longer produce thought, and all thinking stops. But on the other hand, James wrote, the mind can also be a function of the brain the way the keys of an organ make music, by opening the various pipes to let the wind escape in various ways. The organ does not produce the wind or the music – it removes the obstacle holding the wind back.

The association between the mind and the brain is a fact. But the interpretation that the brain *creates* the mind is not a scientific fact. It's only a theory developed to explain the association. And for everyday life, that is a workable model – it's convenient to act as if our brains *do* create our minds. But there are additional scientific findings that suggest there's more to the story. It turns out that the connection between mind and brain breaks down under exceptional circumstances, like near-death experiences.

When the heart stops, breathing also stops, and blood carrying oxygen and fuel no longer flows to the brain. Within ten to twenty seconds, there is no detectable electrical activity in the brain. The person is then considered clinically dead. People who survive such a crisis generally don't have clear thoughts and perceptions during the period when their hearts stopped. And after they are revived, they have no memories of the time they were unconscious. And yet, 10 to 20 percent of such people remember vivid and detailed NDEs that occurred while their hearts were stopped, and some experiencers accurately report events that occurred during that time.

If the mind were in fact produced by electrical and chemical changes in the brain, then near-death experiences that happen when the brain is not functioning should be impossible. If the mind is completely dependent on the brain, how could you have an NDE? How could you have vivid and even heightened feelings, thinking, and memory formation when your heart has stopped and your brain activity is largely gone? NDEs during cardiac arrest and deep anesthesia, when the brain is not able to process experiences and form memories, led me to seek some alternative to the idea that 'the mind is what the brain does.' Again, the furniture in my worldview was starting to fall over. If the brain is not the source of all our thoughts and feelings, then how *do* we explain what's going on with NDEs?

The Mind Is Not the Brain

Near-death experiences raise difficult questions about our understanding of the mind and the brain – an understanding that is already unconvincing. But listening carefully to what experiencers say about their thinking during NDEs may provide some clues to how brain and mind interact.

Steve Luiting had an NDE when he nearly drowned at the age of eight. He was swimming at a local lake on a beautiful, sunny day. He was sunbathing on the beach and without realizing it he developed a very serious sunburn. He then decided to swim out to a raft in the middle of the lake. His friend was doing 'cannonballs,' tucking into a ball to create a giant splash off the raft. Steve decided to try doing one – his first time. He didn't realize that his sunburn had gotten more sensitive as time passed. He dived, tucked into a ball, but then panicked. He straightened out, hitting the water flat on his badly sunburned back. The shock and the searing pain knocked the wind out of him, and he sank under the water, unable to move. He described for me the sensation of his mind expanding beyond the limits of his eight-year-old brain:

'As I continued to sink, I tried moving, but couldn't. I was panicked. As the water grew colder, near the bottom, the pain lessened. I started breathing small amounts of water, thinking that maybe it was possible to get air from water that way. Then, as a part of me realized I was about to die, I began shouting to myself over and over to do something – anything.

'Then, this subtle shift happened. I seemed to change points of view, as if I was changing location in a room. One second, I was the terrified person; and then I was the other calm one "watching" the terrified one. I was both and yet not. The "real" I was the calm one, but I had always identified myself as the other until now.

'My mind expanded to that of an adult capacity, and then beyond. I suppose, without the limitation of a child brain, it allowed my true nature to express itself again. It's made me think that *our understanding of the brain is actually backwards. The brain filters out everything and doesn't help our thinking but hinders it, slows it down, focuses it.* Maybe, because it is so good at filtering and focusing, we don't remember our prior existence – or future events, either.'

Steve described his brain as 'filtering out' and 'focusing' his thoughts, and his mind 'expanding' without the limitation of his brain. And he was not alone in this interpretation of the mind and brain in NDEs. Michele Brown-Ramirez had a near-death experience when she smashed her head on a diving board when she was seventeen. She had always loved watching the high-school diving team, their gracefulness in the air. She eventually joined the team and developed a particular knack for the 'inward dive' – facing away from the pool, toes on the edge of the board, then jumping up, doing a jackknife, and diving into the pool toward the direction of the board. One time during her junior year, she hit the back of her head going down. When she hit the water, she heard everyone on the side of the pool yell in horror. She doesn't know how long she was underwater, but it felt like a long time. Finally, she heard the coach ask someone, 'Do you think I ought to go in and get her?' Like Steve, she described for me the sensation of feeling free of the limitations of her brain:

'At that point, thinking was extraordinarily different, and I remember I could not have breathed even if I wanted to. I "saw stars" everywhere and gradually felt time go faster and slower at the same time, until it felt timeless. I felt a strange pull away from my body and realized I was dying.

'The pull was very strong, and I felt surrounded by presences, people who knew me and each other, but especially my two grandmothers. In that timeless span, I felt free and at peace. It was such a wonderful feeling, I felt like I could "fly" towards a great Light that was God, and a future where I was loved and things made such profound sense. It was a realm of love, peace, calm, and acceptance, which had no space and yet was all space. It was so nice and supportive, and contrary to people here in this world.

'After my head hit the diving board and I was in the water, my thinking was all over the map. It was like a pinball machine on "tilt," and yet there was still some part of me that felt free from ordinary thinking restraints. *I felt free from my brain!* And that "thinking" I had was very free, simple, and clear. It was remarkable to be going through a brain over-firing, or randomly firing, or down, or whatever, and yet to still be able to have free-clear thoughts, and suddenly feel this pull as if I was no longer being constrained to this world and its limits.'

These two NDEs, and many more like them, suggest that your *mind* – the part of you that experiences consciousness – is not the same as your *brain* – the mass of pink-gray matter inside your skull. These experiencers claim that, in their NDEs, their minds were free of the usual limits of consciousness that are present when their brains are functioning normally.

Anita Moorjani, who had an NDE when her body, racked with lymphoma tumors, shut down, explained how the brain limits our awareness of the world around us with the following analogy:

'Imagine, if you will, a huge, dark warehouse. You live there with only one flashlight to see by. Everything you know about what's contained within this enormous space is what you've seen by the beam of one small flashlight. Whenever you want to look for something, you may or may not find it, but that doesn't mean the thing doesn't exist. It's there, but you just haven't shone your light on it.

And even when you do, the object you see may be difficult to make out. You may get a fairly clear idea of it, but often you're left wondering. You can only see what your light is focused on, and only identify that which you already know.

'That is what physical life is like. We're only aware of what we focus our senses on at any given time, and we can only understand what is already familiar.

'Next, imagine that one day, someone flicks on a switch. There for the first time, in a sudden burst of brilliance and sound and color, you can see the entire warehouse, and it's nothing like anything you'd ever imagined . . . You see colors you don't recognize, ones you've never seen before . . .

'The vastness, complexity, depth, and breadth of everything going on around you is almost overwhelming. You can't see all the way to the end of the space, and you know there's more to it than what you can take in from this torrent that's tantalizing your senses and emotions. But you do get a strong feeling that you're actually part of . . . a large and unfolding tapestry that goes beyond sight and sound.

'You understand that what you used to think was your reality was, in fact, hardly a speck within the vast wonder that surrounds you. You can see how all the various parts are interrelated, how they all play off each other, how everything fits. You notice just how many different things there are in the warehouse that you'd never seen, never even dreamed of existing in such splendor and glory of color . . . but here they are, along with everything you already knew. And even the objects you were aware of have an entirely new context so that they, too, seem completely new and strangely superreal.

'Even when the switch goes back off, nothing can take away your understanding and clarity, the wonder and beauty, or the fabulous aliveness of the experience. Nothing can ever cancel your knowledge of all that exists in the warehouse. You're now far more aware of what's there, how to access it, and what's possible than you ever were with your little flashlight. And you're left with a sense of awe over everything you experienced in those blindingly lucid moments. Life has taken on a different meaning, and your new experiences moving forward are created from this awareness.'

The idea that our minds are independent of our brains seems to fly in the face of our everyday experience. Doesn't the brain do our thinking? That is, isn't the mind simply 'what the brain does'? And yet the accounts of these near-death experiences from people like Steve, Michele, and Anita made me consider seriously the idea that our everyday experience is not the whole story, and that our minds can sometimes act independently of our brains. In the face of challenging evidence like these accounts, I needed to explore whether an alternative model of the brain-mind relationship was possible.

And in fact, that is what an increasing number of scientists are doing. More than a decade ago, I participated in a symposium at the United Nations on alternative models for the mind and brain. Since then, a survey of two hundred fifty Scottish university students from eight different fields (86 percent of them various sciences) found that two-thirds believed that the mind and the brain are two separate things. A similar survey of almost two thousand Belgian medical professionals found that the majority believed the mind and brain are two separate things. And a recent survey of more than six hundred Brazilian psychiatrists found that the majority believed the mind was independent of the brain. An increasing number of scientists around the world are finding that the old model – the mind being totally dependent on the brain – is inadequate.

The concept that the brain produces the mind is a reasonably good guide to everyday life. When our brains are damaged by a hard blow or by a viral infection or by drinking too much, we don't think as well. But that mind-brain model falls down when we get to some extreme conditions, particularly when the brain stops working but the mind keeps going. Basil Eldadah, supervisory medical officer at the National Institute on Aging, recently wrote that 'dominant paradigms unwittingly create barriers that hinder innovation . . . While prevailing theories tend to do best at explaining averages, they can break down at the extremes. And if a theory cannot adequately explain the extremes, then either the extremes are an artifact or the theory needs a second look.'

It may help to think of our models as tools for dealing with the world. We need different tools for different tasks. A hammer is an excellent tool for driving a nail into a piece of wood, but it's not the best tool for screwing a nut onto a bolt. Although I do find hammers very useful for driving nails, I wouldn't insist that they're the only tool worth using for any task. In the same way, the idea that our brain creates our thoughts and feelings is a useful tool for every-day life, but it's not the best tool for understanding NDEs, when thoughts and senses become more vivid than ever even though the brain is deprived of oxygen and fuel.

But what's a better way to think about the relationship between the chemical and electrical activity in your brain and the thoughts and feelings in your mind? One answer is that the brain is a device for the mind to act more effect-ively on the physical body, to focus our thoughts on the physical world. As French philosopher Henri Bergson put it, 'the brain maintains consciousness fixed on the world in which we live; it is the organ of attention to life.' That is, your brain can receive thoughts from your mind, select the ones that are important for your survival, and convert them into electrical and chemical signals that your body can understand.

Your mind has many thoughts that have nothing to do with surviving in the physical world. Think of all the unusual things that people describe happening in NDEs, from meeting guides and deceased loved ones to visiting otherworldly places. Thoughts and feelings like those don't help us survive in the physical world, and may in fact get in the way of our ability to process information about the world quickly. So the brain works like a filter to block out information the body doesn't need for survival, and selects from the thoughts and memories stored in the mind only the information that the body needs.

It's very much like the way a radio receiver selects from

the variety of broadcasts only the signal we want to listen to, and filters out all the other stations. If it didn't do that, our ears would be overwhelmed by trying to listen to hundreds of radio stations at the same time, and we wouldn't be able to hear any of them clearly. But when the brain is impaired, as in a blow to the head or during anesthesia or intoxication, its ability to filter our thoughts and feelings goes on the blink. It's like the garbled sounds you hear when you adjust the radio's tuning knob between stations.

This explains why some experiencers say that, once returned to their physical bodies after an NDE, their minds were once again tied to their brains and they could no longer understand things as they could when they were free of the constraints of the human brain. Lynn had an NDE when she was hit head-on by a drunk driver while riding her bike at age twenty-one. She told me that once she woke up back in her physical body, she could no longer understand things that were obvious during her NDE:

'I was riding home from a meeting and I had a head-on collision with a girl who had run through a red light. There happened to be a nurse driving behind me who had seen me at the meeting and knew who I was by my helmet. She had come up behind me a little while after I had wrecked, and she saved my life. Then a rescue squad came later and helped, but she started cardiopulmonary resuscitation on me before they came. I don't remember anything about my accident, or even the time around it. That's why I think it is weird that I remember what I do about going to that place.

'The first thing I remember is that everything was black. I remember seeing this light and then towards the end, though I still had pretty far to go, I came upon it faster than I had gone through the black. I guess that was where the time ended, because I really wanted to get to the light and I couldn't get there fast enough, and the distance was time. The black was huge, and the light – I am not sure they ever ended. I remember knowing that back on earth nobody could comprehend this.

'It wasn't scary moving through the black. I was just watching. Then there was this light at the end. It wasn't just a light, like a transparent color: it was intense love. When you are in it, it doesn't just surround you like water does when you jump into a pool. It was like sun going through a piece of glass. It went completely through every spot on your body, and everywhere else. It was warm and comforting. It was like warmth, comforting, a peaceful silence, and love surrounding everything and within.

'There were no walls or boundaries or anything solid, just light and beings. The light was like a magnet, too. You just cannot be apart from it; you want to be with it more than anything you've ever wanted. Everyone loved each other more than can be comprehended here, because of what we were, not who we were. We're limited, but they are not. I don't know how to explain how we talked. We didn't talk like we do here. We just knew. There was no time at all, not from one second to the next event. I'm not sure how to explain it. It's like you are wholly present, with no memory of past, and no future exists. I had no body, only sight. After that, I just went flying backwards really fast through the black. I remember finally, it was like I had sound, but no brain.

'The next thing I remember was just lying there. I could hear this beeping from the heart monitor, and my sister screaming in my ear, "Lynn, this is Caroline, your sister! You were in an accident!" I couldn't respond or move. About a week and a half after my accident, I still didn't know who anybody was, and I couldn't open my eyes. Later, even after I opened my eyes, I still didn't know who anybody was, or who I was.

'I remember, finally, about a month later, my brain started to work a little bit. I remember opening my eyes and all I could think about was, "Shit! All I have is a human brain." I said that, too, and was complaining about being a person again, and I know my mom and my sister thought I was nuts.

'I also remember, when I got back, knowing that *as long as I am on earth, I will never be able to comprehend it, either, because I only have a human brain.* Here we can really think about only one thing at a time, and there you know – really know – everything. You can't compare it to earth things. Talking about it or trying to draw it diminishes it entirely. It would be like trying to talk to an infant about DNA or some kind of medical technology in space. An infant couldn't even speak the language, and he definitely couldn't understand the idea. He could know about things only on his level, like we do. We are just like babies, and contrary to what a lot of people think, we know nothing. I'll never be able to feel what I felt there, while I'm here, because I'm back in this human body again. It is way beyond, superior, bigger than anything a human brain can comprehend, and more wonderful, too. But I think it is like a party: you cannot go until you get invited. I feel like an ant in an ant farm.'

The idea that the brain processes or filters our thoughts rather than creating them is not new. It's been described over the centuries with various metaphors. The Greek physician Hippocrates wrote about this model more than two thousand years ago: 'The brain is the most powerful organ of the human body, for when it is healthy it is an interpreter to us of the phenomena . . . that gives it intelligence . . . To consciousness the brain is the messenger . . . Wherefore I assert that the brain is the interpreter of consciousness.'

English philosopher Aldous Huxley described this model using the metaphor of the technology of the last century:

'The function of the brain and nervous system is to protect us from being overwhelmed and confused by this mass of largely useless and irrelevant knowledge, by shutting out most of what we should otherwise perceive or remember at any moment, and leaving only that very small and special selection which is likely to be practically useful . . . In so far as we are animals, our business is at all costs to survive. To make biological survival possible, Mind at Large has to be funneled through the reducing valve of the brain and nervous system. What comes out at the other end is a measly trickle of the kind of consciousness which will help us to stay alive on the surface of this particular planet.'

If I were to call you on your cell phone, you'd hear my voice coming out of your phone, but you wouldn't think that the cell phone was *creating* my voice all by itself. You'd know that my voice was coming from me, and that radio waves carry my voice to your phone, which then re-creates the sound of my voice for you to hear. If your cell phone was damaged or ran out of power, you'd no longer be able to hear my voice. I'd still be able to speak, but you'd no longer be able to hear my voice through your cell phone. Your brain may function like a cell phone. It receives the thoughts and feelings and converts them into electrical and chemical signals that the body can understand and use.

The idea that the brain is a filter, limiting incoming information to what's important to our physical survival, should not be surprising. All our senses filter out unimportant input. Our eyes not only transmit light to us, but also filter out ultraviolet and infrared light, so that we see only the small portion of light that is in our visible range. Remember Jayne Smith's comment that in her NDE during a bad reaction to anesthesia she saw 'flowers that had colors that I'd never seen before. And I remember so well looking at them and thinking, "I have never seen some of these colors!"' Likewise, our ears not only receive sound vibrations but filter out most of the sound frequencies that are important to dogs and cats but not to us. So it is consistent with what we know of our neurobiology that, if our thoughts and feelings come from outside the body, the brain would act to filter out those that are not essential to our physical survival, just as other parts of our nervous system filter out nonessential information coming from the outside.

The suggestion that the mind could operate independently of the physical brain seems counterintuitive, but it is not outside the reach of science. Currently neuroscientists are working out possible biological mechanisms by which the brain could act as a filter, focusing primarily on the prefrontal cortex, which controls selective attention, and on electrical activity that is synchronized across different parts of the brain.

I wondered whether there was evidence for the filter model of the brain beyond what I was finding in my research on NDEs. What I discovered is that there is quite a bit. A similar unexplained experience is something called 'terminal lucidity' or 'paradoxical lucidity,' in which someone who has had an irreversible brain disease for years, like Alzheimer's disease, and is unable to speak or to recognize family, suddenly becomes mentally clear again. People who have terminal lucidity regain the ability to recognize family and carry on meaningful conversations and express appropriate emotions, for no apparent neurological reason.

This astonishing and unexplained recovery usually happens in the hours before the person dies, suggesting that the deteriorating brain has lost its ability to filter the mind, which is briefly free to express itself before the person dies. Terminal lucidity is extremely rare, but the fact that it happens at all is a puzzle for neuroscientists. A couple of years ago I participated in a workshop at the National Institute on Aging to assess what we know about terminal lucidity and to identify areas ripe for further study. That workshop resulted in the institute releasing two funding opportunity announcements to foster research on this sudden, unexplained clear thinking in advanced Alzheimer's disease.

In addition, recent neuroimaging studies of people under the influence of psychedelic drugs have shown that the elaborate mystical experiences associated with these drugs are accompanied by *decreased* brain activity. This is the exact opposite of what we had expected. Traditional neuroscientific explanations of psychedelic drug trips assumed that psychedelic drugs like LSD and psilocybin increase brain activity, triggering hallucinations. But it appears that they instead decrease brain activity, particularly in the prefrontal cortex, and also sharply decrease the kind of synchronized electrical activity in the brain that is typically seen with complex thinking. This decrease in brain activity might reduce the brain's ability to filter the mind and permit access to mystical experiences. This is consistent with spiritual traditions from around the world that use choking, breath-holding, starvation, and extended sensory deprivation to induce mystical experiences.

These studies suggest that profound experiences may be associated with decreased brain activity and decreased connectivity between different brain regions. NDEs may be the ultimate example of elaborate experiences associated with not only reduced, but practically absent brain activity. All this evidence is consistent with the idea that the brain is a filter of our thoughts and feelings, and that the range of our thoughts and feelings expands as the filtering activity of the brain decreases. As physician Larry Dossey put it, 'We are conscious not *because* of the brain but *in spite of* it.'

If our minds can function independently of our brains, might that allow us to make some sense of the more puzzling features of NDEs? Might it help us understand how Al Sullivan 'saw' his surgeon flapping his arms, how Bill Hernlund 'saw' his colleagues drag his body away, and how Tom Sawyer was able to review his entire life in minute detail?

And if our minds can function even when our brains are offline, as this evidence suggests, then might it be possible that our minds could continue to function after our brains have stopped permanently – that is, after we die? This question seemed to be beyond the traditional realm of science – and yet scientists around the world have been increasingly willing to stretch the boundaries of that traditional realm in recent decades. This question of whether our minds might survive death also challenged my personal view of how the world works. Nothing in my background or training had prepared me to take that possibility seriously. But as I discovered, there are ways to bring scientific principles and methods to the question of whether our consciousness might be able to continue beyond death.

Does Consciousness Continue?

Almost without exception, people who have had near-death experiences hold a firm belief that some part of them will live on after death. No matter what they think will happen after their bodies die, they do not think physical death will be the end of them. And although their ideas about exactly *what* might happen after death differ from one person to another, there are some recurring patterns in their descriptions of an afterlife existence. For example, three-fourths of the experiencers who participated in my research said that the afterlife is *a blissful state of peace and tranquility*, with no pain or suffering. They also described the afterlife as *outside of time*, and said that earthly time as we know it no longer existed for them in that realm. Two-thirds of the experiencers said that we continue to exist in some identifiable form with our own thoughts, feelings, and personality traits, but that we continue to learn and grow spiritually after death.

More than half of those who had NDEs said that after death we review our lives and judge ourselves, and in some form have to face the consequences of things we did in life. They also said that what happens to us in the afterlife is at least partly dependent on how we lived before death, and that we will also reap benefits for our good deeds and actions in this life.

Almost half of the experiencers said that in the afterlife we will still be able to watch loved ones who are still alive, and may be able to communicate and interact with them. They also said that we still have physical-like sensations in the afterlife, comparable to seeing and hearing, and that we still have emotions.

Two-thirds said that in the afterlife, we meet loved ones who had died earlier, and that in fact they did see or sense the presence of a deceased loved one in their NDE. Such experiences are convincing to the experiencers themselves, but do they provide enough evidence to convince the rest of us that death is not the end? Can these reports of meeting deceased loved ones be verified, or are they just wishful thinking, reflections of our expectations and hopes of what might happen when we approach death? It is possible, of course, that at least some of those reports stem from the experiencers' imaginations, and from their hope to be reunited with loved ones after death. But there are other NDEs that can't be written off as expectation. Sometimes experiencers meet recently deceased people *who were not known to have died.*

Jack Bybee was hospitalized with severe pneumonia at age twenty-six, in his native South Africa. He described for me an encounter with his nurse during his NDE:

'I had been taken very ill, and was three to four weeks in an oxygen tent in status epilepticus, then double pneumonia, and so on and so on. I was friendly (read "flirting," when I could be) with a nurse from the farmlands of the Western Cape. She had told me it was her twenty-first birthday that weekend, and that her parents were coming in from the country to celebrate. She fluffed up my pillows, as she always did. I held her hand to wish her a happy birthday, and she left.

'In my NDE, I met Nurse Anita on the other side. "What are you doing here, Anita?" I asked. "Why, Jack, I've come to fluff up your pillows, of course, and to see that you are all right. But, Jack, you must return, go back. Tell my parents I'm sorry I wrecked the red MGB. Tell them I love them."

'Then Anita was gone – gone through and over a very green valley and through a fence, where, she told me, "there is a garden on the other side. But you cannot see it. For you must return, while I continue through the gate."

'When I recovered, I told a nurse what Anita had said. This girl burst out into tears and fled the ward. I later learned that Anita and this nurse had been great friends. Anita had been surprised by her parents, who loved her dearly and had presented her with a red MGB sports car. Anita had jumped into the car, and in her excitement raced down the highway, De Waal Drive, along the slopes of Table Mountain, into "Suicide Corner" and a concrete telephone pole.

'But I was "dead" when all that happened. How could I possibly know these facts? I knew them as stated above. I was told by Anita in my experience.'

When Jack told me this story about fifteen years ago, emphasizing his astonishment at meeting the nurse he thought was still alive, I recognized right away that there was something important in his account I hadn't appreciated before. I *had* heard many accounts of experiencers meeting deceased loved ones in their NDEs, starting with Henry more than thirty years earlier telling me he saw his parents after he shot himself. Of course, Henry knew that his parents had died and, what's more, yearned to see them again, which made me as a young psychiatrist suspect he had hallucinated them. But Jack had no way of knowing that his nurse had died, and no yearning to see her on her weekend off with her parents. This was an apparent encounter with a deceased person that couldn't be dismissed as wishful thinking. And Jack was not the only experiencer who told me a story like this.

Later that same year, one-hundred-year-old Rose described for me a comparable NDE when she was hospitalized with pneumonia during World War I:

'During the war, I was very ill in the hospital. One morning the nurse came in and found me showing no sign of life whatever. She called the doctors and the matron, to whom I also appeared dead, and I remained so, they told me afterwards, for at least twenty minutes.

'Meanwhile, I found myself in a beautiful, green,

undulating country. There were beautiful, large trees here and there, and their leaves seemed to give out a sort of gentle radiance. I then saw a young officer with a few soldiers approaching. The young officer was my favorite cousin, Alban. I did not know that he was "dead," nor had I ever seen him in uniform; but what I saw of this was confirmed by a photograph of him I saw some years later.

'We spoke for a few minutes happily and then he and the few men marched off. Then a presence beside me explained to me that all of these soldiers were allowed to go and greet and help those others who met death on the battlefields.

'My next vivid recollection after this was of looking down, from about ceiling height, onto a bed on which lay a very emaciated body. There were white-coated doctors and nurses around it. In a few moments I was looking up at them, and feeling a sensation of intense disappointment. I had come back from something so lovely and so utterly satisfying.'

And a few years later, Barbara Langer told me about a similar NDE she had had after a car crash at age twenty-three:

'I was recuperating from hepatitis at the home of a young couple I'd recently come to know, David and Christine. Christine soon became a close friend and took care of me. I'd been staying in their guest room for a few weeks when the accident happened.

'Christine and I were taking their sick white cat to see the veterinarian on a sunny Tuesday afternoon. She was driving David's Volkswagen bus into town and we were talking about an upcoming concert. I was holding the cat, Nasty, on my lap. Suddenly Nasty wriggled free. He jumped onto Christine's arm and began climbing up her neck. She tried to push him off and I reached over and tried to get him off her. That's the last thing I remember about our intended trip to see the vet. Later, I learned that we'd crashed into the back of a school bus and both of us went through the windshield. I was unconscious for almost a week.

'I felt that I was hurtling through a dark and vast universe at a tremendous speed. I felt small, calm, and detached but interested in my trip, while at the same time I was very aware of the speed at which I was traveling. This seemed to go on for a long time.

'I found myself blissfully alone in a very green meadow, in a valley surrounded by rounded green hills. There were flowers, and a stream there. I remember the colors were beautiful and intense and the atmosphere was ethereal. The feeling of all-pervading peace and love that I felt was gloriously uplifting.

'Then I found myself going down a path in another lofty, ephemeral place. Christine was next to me and wearing the same blue jeans that she'd put on that day, and we were gliding down the pathway side by side. She looked serene and astoundingly beautiful. I have no idea of how I appeared, but I knew that the two of us were surrounded by and filled with love.

'We were on a narrow, earthen footpath. Soon the pathway split, going into two different directions. We both knew that this was the place where we would have to part and go our separate ways. We were calm and communicated without words. Christine telepathically asked me to be sure to tell David that she would love him forever. She then took the path that went right; I took the path that went left. It was unclear as to where these pathways would lead us. There was no conscious decision on our parts as to what we should do, but it was clear which way each of us needed to go. We parted with the shared understanding that we would be together again, but later. Now, however, we'd each have to undertake the hard part of the journey that lay ahead of us without each other's company.

'My path led me instantaneously back to my physical body. I regained consciousness in the hospital. There were still slivers of window glass from the car windshield in my hand, and a gash on my forehead. I didn't recognize my face in a mirror that was held up for me, and was told by a friend that I'd just gotten "a face lift from the middle." I had double vision from the triple concussion the doctors said I'd gotten, but I hadn't lost my contact lenses and my hepatitis was gone.

'Friends and family came to visit. One friend showed me a newspaper article about the accident, and that's when I learned about Christine's death. Christine reportedly died at the scene of the crash, and I was taken to a nearby hospital and not expected to live.

'Putting the various pieces of my experience together, I realized that I had died and come back and my friend didn't, but she was in a much nicer place. I wished that I had stayed there with her because there was nothing in my life then that made me feel invested in the physical world.'

When Jack, Rose, and Barbara described to me their NDEs, I recognized that these were important experiences to investigate. Meeting deceased loved ones is not unusual in NDEs. Of all the experiencers who have participated in my research, almost half report meeting someone who had died. I no longer automatically jump to the conclusion that such accounts are hallucinations. But I also don't regard most of them as scientific evidence of anything, because I can't rule out the influence of the experiencers' hopes and expectations of meeting their loved ones. However, when it comes to meetings with people who were not known to have died, such as Jack's, Rose's, and Barbara's, I knew they couldn't be dismissed as expectations of a reunion. But I kept looking for other possible explanations.

Could these visions have been made up after the fact? That is, might experiencers meet some being in their NDEs, and only after learning that a loved one had just died, retroactively identify the being they met as that newly deceased loved one? That may happen in some NDEs, but in other cases, like Jack's, the experiencer told other people about the vision, naming the deceased person, before learning of his or her death.

Could these visions be 'lucky guesses'? That is, might experiencers envision meeting someone in their NDEs who was still living but likely to have died? If that was the

explanation, then were there also NDEs where experiencers 'guessed wrong' and identified people in their NDEs as
deceased who were really still alive? It turns out that there
are a few NDEs in which experiencers report meeting
people who are still alive. In our collection that now
includes more than a thousand NDEs, 7 percent involved
seeing someone in the realm of the NDE who was still living. But in every one of those rare cases, the experiencer
described that person as still living, in most of those cases
pleading with the experiencer to come back. *None* of the
NDEs in our collection involved an experiencer mistak-
enly thinking a person still alive had died.

NDEs in which the experiencer meets – and is surprised
to see – a loved one they hadn't known had died are not
common, but they do occur. And cases of this type are by
no means new. They've also been recorded throughout the
ages.

The Roman historian and naturalist Pliny the Elder
wrote in the first century about a nobleman named Corfidius, who was pronounced dead by a physician when he
stopped breathing. His will was opened, naming his younger
brother as his executor and heir. The younger brother then
hired an undertaker to arrange the funeral. Corfidius, however, stunned the undertaker by sitting up on the embalming
table and clapping his hands, a typical signal to summon his
servants. He then announced that he had just come from
the home of his younger brother. He said that his brother
told him that the funeral arrangements he'd made for

Corfidius should now be used for him instead. He said his brother also asked Corfidius to take care of his daughter, and showed Corfidius where he had secretly buried some gold in his yard. As Corfidius was relating his story to the astonished undertaker, his younger brother's servant burst in with the news that his master had just unexpectedly died. The buried gold was indeed found where Corfidius said his brother had indicated.

Several detailed and well-documented cases of this type were published in the nineteenth century. Physicist Eleanor Sidgwick wrote of an Englishwoman who was seeking a singing instructor for her visiting nieces. She hired Julia, the daughter of a local tradesman, who'd been trained as a professional singer. After the nieces had gone, Julia told her father that she'd never had a happier week. Shortly thereafter, Julia married and moved away. Several years later, the woman who had hired Julia lay dying and was going over some business matters when she suddenly stopped and asked, 'Do you hear those voices singing?' No one else in the room did, and she concluded that they must be angels welcoming her into heaven, adding, 'But it is strange, there is one voice amongst them I am sure I know, and cannot remember whose voice it is.' Suddenly she pointed up and said, 'Why, there she is in the corner of the room. It is Julia.' No one else saw the vision, and the woman died the next day, February 13, 1874. On February 14, Julia's death was announced in the London *Times*. When her father was later interviewed, he reported, 'On the day she died she began singing in the morning, and sang and sang until she died.'

More recently, Dr K. M. Dale reported the case of nine-year-old Eddie, whose fever finally broke after nearly thirty-six hours of anxious vigil on the part of his parents and hospital staff. As soon as he opened his eyes at three a.m., Eddie urgently told his parents that he'd been to heaven, where he saw his deceased grandpa, Auntie Rosa, and Uncle Lorenzo. His father was embarrassed that the doctor was overhearing Eddie's story and dismissed it as a feverish delirium. Then Eddie added that he'd also seen his nineteen-year-old sister, Teresa, who told him he had to go back. His father then became upset and asked Dr Dale to sedate Eddie. He had just spoken with Teresa, who was attending college in another state six hundred miles away, two nights earlier. Later that morning, when his parents called the college, they learned that Teresa had been killed in a car accident just after midnight, and that college officials had been trying to reach them at their home.

Are these visions nothing more than fantasies, or can we really meet deceased loved ones in our NDEs? Having had no religious belief in an afterlife to fall back on, I found these visions of deceased people who were not known to have died to be very difficult – if not impossible – to explain. Experiencers report that these deceased people not only appeared to them but *interacted* with them, giving them information. Who or what was it that was giving the experiencers this information? The experiencers' interpretation in every case was that it was the deceased person, who was still somehow conscious and able to interact. But that would require that consciousness – the ability to think and feel – continues after the physical body dies. I had a hard time understanding how that could be. If our consciousness doesn't end when our bodies die, then where does it go?

. 13

....

Heaven or Hell?

Questions like 'Where does consciousness go after the body dies?' seem to be pushing the limits of scientific research beyond the breaking point. If we look to what experiencers tell us about where they go when they leave their bodies, many of them provide elaborate descriptions of the kind of condition in which they find themselves. For most of them, their NDEs take them to a world of bliss.

Among all the experiencers in my research, almost 90 percent reported feelings of peace. Almost three-fourths reported feelings of joy or bliss. Two-thirds described a heavenly sense of cosmic unity or being 'one with everything.' And three-fourths reported an encounter with 'a loving being of light.' I was surprised when I first started studying people who have NDEs to find that for the most part they are *not* terrified and panicking as they face death.

In fact, it is usually just the opposite. Most people who report NDEs describe overwhelmingly positive feelings, ranging from incredible calmness to joy and ecstasy.

Is that what people through the ages have meant when they described 'heaven'? People often describe their NDEs as occurring in some realm so different from our everyday physical world that they feel that our words are not adequate in describing them. And because they find it difficult, if not impossible, to describe these other realms and beings accurately, they often resort to whatever cultural or personal metaphors they have available to describe things that don't seem to fit familiar labels – and one of the most familiar labels is 'heaven.'

Many experiencers identify the blissful other world in which they found themselves literally as 'heaven.' Judy Friel, who was raised in a Presbyterian family 'when we attended church,' felt herself rise above her hospital bed during a medical crisis at age twenty-four. She described for me where she went:

'I floated to heaven. I knew this was heaven because of what the Bible says heaven is like, and because of teachings at church. I liked heaven. I and everybody else were at peace, happy, and felt no pain. I was met by an angel who told me to come in and see what heaven was like. I saw people working. They were singing and laughing. Young and old were together. The music was beautiful and in harmony, even from those who could not sing. I recognized some people. I saw people as I saw them on earth, the same age and with the same clothes on. And yet, at the same time, all were alike, wearing the purest white robes I'd ever seen – actually, whiter than that. I saw streets and rows of mansions. All glittered in fine gold.

'I found myself standing in a long line. After a short while, I realized I was in a line leading to the throne of God. I was going to be asked to give an account of my life. The throne was surrounded by a brilliant white light.'

Dottie Bush also described visiting a place she identified as heaven when she hemorrhaged during childbirth at age twenty-five:

'I had gone into shock. The last thing I remember was the anesthetist yelling to the doctor to hurry, that my blood pressure was falling.

'I then found myself in a beautiful place. I know it was heaven: so peace-filled, so beautiful, and such lovely music and flowers. They were *so* beautiful that they seemed many times more beautiful than the ones we see here on earth. With *that* kind of music and being surrounded by love and peace, I did not want to return.

'Then someone started to talk to me. I didn't see his face, but I felt it was Jesus. He said, "Dottie, I am leaving you on earth for a purpose." And He proceeded to make known to me *all things*. He told me why he died on the cross, but I only remember that it was *a bit different* than what has been taught by the church.

'I felt, as Jesus talked to me, "Why did He choose me to reveal all things to?" Then I thought that, since He did, now that I have had this convincing experience, I can be of help to others by helping them to understand. When He finished talking to me, I felt myself floating away from that beautiful place to this dirty and ugly one, so great was the contrast between heaven and earth. I did not want to return, although He said that I must.

'I then felt myself back in my body on the operating table. My doctor indicated resuscitation measures had been required because of apnea and hypoxia.'

Judy and Dottie both identified the 'someplace else' they went as heaven. Most experiencers, however, are not able to identify where they went, but simply describe the other world in which they find themselves without putting a label on it. Three-fourths of the experiencers in my research described entering some unfamiliar realm or dimension. Although most said that other place could not be described in words, when pressed to try to describe it, they used a variety of metaphors, including religious terms like heaven or hell or natural terms like a valley or meadow or 'outer space.' But even when pressured, almost half still insisted they could not find *any* familiar label that fit it. Cynthia Ploski, who had been raised Protestant but later in life regarded herself as spiritual but without any religious affiliation, had a heart attack at age seventy-two. She described a blissful realm in natural terms:

'As for the moment my heart stopped, I was not aware of it. I was suddenly, without any sense of feeling death, standing at the edge of this beautiful woods. There were no people there, only soft, golden light, spring green leaves around, and above me, gentle breezes rustling the leaves. It was like being in a vision, but very real and vivid, *unmistakably real*. I had a sense of standing there, but do not know what form my body had, only that I was not floating or wispy. I felt solid and normal. And I was capable of cognitive thought, because I realized where I was, on the "other side," and knew I did not want to stay there.

'Ahead of me, a sort of glade or path led deeper into the woods, and there was more light down at the end. It seemed inviting to enter, but I thought that if I didn't get out of there I wouldn't be able to get back to the life on this side. So I thought, "I'd better get out of here!" and with that thought I was back in the emergency room.'

Likewise, Harriet, who had an NDE at age seventy-four during a heart attack, described a blissful experience:

'I seemed to be floating in a more or less confined space, but there were no walls as we know them. I was moving in and out of a billowing, soft, dark purple, velvety substance. It was beautiful, sensual, voluptuous, sort of like falling into a great mass of soft satin and down feathers. I was completely surrounded by this substance, and I floated up and down slowly, restfully.

'Each time I got near the bottom, I could see a great brightness at the end of this space, slightly to the right. This brightness was warm, soft, and so welcoming. I floated down to it a few times, but even though I got close, I never made any effort of my own to reach it. I didn't seem to have a body or mind. I didn't seem to be a person, or even a thing. I was peaceful, happy, contented. I didn't seem to care about anything anymore. It is not a feeling you can put into words: no mind, no body, no boundaries, only contentment, sort of like an amoeba that had gotten into the ocean by mistake.'

Unlike Judy and Barbara, Cynthia and Harriet did not label the world they went to but simply described it as 'a beautiful woods' and 'a billowing, soft, dark purple, velvety substance.'

But not all NDEs are described as blissful or pleasant. When I first started investigating near-death experiences in the late 1970s, the accounts I heard were overwhelmingly peaceful, if not blissful. In the intervening years, however, I have found that, although *most* people describe pleasant feelings in their NDEs, there are *some* who do not. By the early 1990s, Nancy Evans Bush and I had collected enough accounts of distressing NDEs to publish the first report of such experiences in a medical journal.

Among our current sample of experiencers, 86 percent said their NDE was primarily pleasant, 8 percent said it was unpleasant, and 6 percent said it was neither. Although only a small minority of experiencers describe frightening or distressing experiences, it is possible that there are many more people who have unpleasant NDEs but are unwilling to talk about them. For that reason, I am not confident that frightening near-death experiences are as rare as they appear to be.

From the hundreds of NDE accounts I have collected, and the investigations of other near-death researchers, there is no obvious reason to explain why some people have blissful NDEs and others have frightening ones. It is not true, for example, that people who live 'saintly' lives always have pleasant experiences while 'bad' people always have terrifying ones. Throughout history, revered mystics like the sixteenth-century Saint Teresa of Ávila and Saint John of the Cross, and the twentieth-century Mother Teresa of Calcutta, described their 'dark night of the soul' as a necessary first stage on the path to union with the Divine.

On the other hand, I have heard accounts of blissful NDEs from career criminals, including murderers serving life sentences in prison. The meager amount of evidence we have from frightening or distressing NDEs at this point suggests that they can occur under the same conditions as do blissful NDEs. We don't know why some people have distressing NDEs while others have blissful ones. I have noticed that reluctance to face frightening NDEs may lead to long-lasting emotional trauma, and that distressing NDEs are often interpreted as a message for the experiencers to turn their lives around.

Some experiencers describe a realm that sounds like the traditional descriptions of hell. Brenda had one such experience at age twenty-six when she attempted suicide by overdosing on sleeping pills:

'The doctor at the hospital leaned over close to me and told me I was dying. The muscles in my body began to jerk upward, out of control. I could no longer speak, but I knew what was happening. Although my body slowed down, things around me and things happening to me went rather fast.

'I then felt my body slipping down – not straight down, but on an angle, as if on a slide. It was cold, dark, and watery. When I reached the bottom, it resembled the entrance to a cave, with what looked like webs hanging. The inside of the cave was gray and brown in color.

'I heard cries, wails, moans, and the gnashing of teeth. I saw these beings that resembled humans, with the shape of a head and body, but they were ugly and grotesque. I remember colors like red, green, and purple, but can't positively remember if these were the colors of these beings. They were frightening and sounded like they were tormented, in agony. No one spoke to me.

'I never went inside the cave, but stood at the entrance only. I remember saying to myself, "I don't want to stay here." I tried to lift myself up, as though trying to pull myself – my spirit – up out of this pit. That's the last thing I remember.'

Brenda recovered from this overdose and began treatment for the depression that had led to her trying to take her life, and started attending Alcoholics Anonymous meetings. Despite the hellish character of her NDE, her new belief that death is not the end gave her hope and motivation to turn her life around, and she eventually became a counselor to people dealing with depression and substance abuse.

Kat Dunkle described her hellish experience at age twenty-six when she had massive internal bleeding from a car accident:

'The accident rendered a severe blow to my back and the membrane surrounding my jellylike liver had exploded like a balloon, causing massive internal bleeding. The considerable loss of blood made a heroic effort by the surgeon almost impossible. I had flatlined and by all medical accounts was clinically dead. The anesthesiologist had turned off the equipment and stood up to leave but the young surgeon would not give up and brought me back to life. While these events were playing out in the cold sterile operating room, I was on a journey that would change my life forever.

'I was clinically dead on the operating table. I felt myself take my last breath, was hurled down a tunnel. Then the floor dropped out and I fell into total darkness with the horrid pain of burning engulfing my entire body. I heard others screaming and I knew I was in hell. I knew there was no escape and that I would fall and burn, screaming into the total darkness forever. I cried out to God to help me but I knew He didn't hear me and that no one even knew I was there. Then it just stopped.

'Instantly I started plummeting downward, falling into darkness, a horrible endless black space. Imagine standing in an elevator and all of a sudden the floor drops out and down you go, that terrible sensation of falling. I was terrified in the darkness that surrounded me but very aware of the horrible pain burning and searing my entire body, agonizing pain beyond description that would never leave. There were the tortured screams of others but I could see nothing but the darkness. There was no fire, just this dreadful burning pain over every part of me and I knew that this was hell.

'I felt hopelessness – knowing this was for eternity! There was no escape from the nightmare: I wouldn't wake up; I wouldn't hit bottom and die; I wouldn't be rescued by anyone. I would fall and burn in this gruesome place forever and ever and ever, screaming out with all these other lost souls crying out in the darkness, totally helpless as we fell further into the pit of hell. Not even God entered into this place and the torture would go on forever and ever and ever. There is no way to describe the terror that filled me, realizing that I actually sent myself to hell through my choice of not believing. I had chosen this. I had chosen not to believe in God.

'I felt a separation, as if I had never existed. There is no lonelier place than separation from God. I saw no flames, just total darkness and the sensation of burning. I heard many people screaming but I saw no one. It was a dark, desolate, horrible place with no hope of escape. I felt the hopelessness of being lost in torment, separated from God for eternity.'

But Kat's experience didn't end there. As with many frightening NDEs, hers eventually turned around and became peaceful. In Kat's case, even though she had been an atheist for her previous twenty-six years, she cried out for help from God:

'As I was falling, burning in this horrifying place, I cried out to God, begging him to forgive me. I pleaded that he would release me from this place. Then the torture stopped. It just stopped! The loud, piercing, howling noise that rang through my ears and the horrendous feeling of burning and tearing through the middle of my body stopped, and I knew without any doubt: "There is a God." I was filled with the perfect peace of God, a peace that cannot ever be described, peace that transcends all understanding. There was no fear, pain, anxiety, or emotions of any kind. Everything was overcome by a feeling of worship towards God and of really knowing Him. So I went from being a total nonbeliever to a person who has no doubt.'

Brenda and Kat clearly identified their 'someplace else' as hell, just as Judy and Dottie identified their 'someplace else' as heaven. Most survivors of frightening NDEs, however, just like survivors of blissful NDEs, do not give a name to where they went, but simply describe the other world in which they find themselves without putting a label on it. Doris described to me a frightening NDE when her cervix and uterus tore during childbirth at age twenty-seven:

'Suddenly I became aware that something really strange was happening. It was as if I had pulled up and away from my body, and I found myself watching my doctor and his nurse working on my body from a corner of the room near the ceiling. I felt so startled at being able to hover above like that, and I wanted to feel in control of my situation, but I was unable to do anything except watch helplessly.

'Then I found myself no longer in the room, but traveling through a tunnel, slowly at first, then picking up speed as I went. As I entered the tunnel, I began hearing the sound of an engine, the kind that operates heavy machinery. Then, as I was moving slowly, I could hear voices on each side of my head, the voices of people whom I've known before, because they were vaguely familiar. About this same time, I became frightened, so I didn't concentrate on trying to recognize any of the voices.

'I found myself growing more and more afraid as the speed picked up, and I realized I was headed toward a pinpoint of light at the end of the tunnel. The thought came that this was probably what it was like to die. I decided then and there that I wanted to go no further, and I tried to backpedal, stop, and turn around, but to no avail. I could control nothing, and the pinpoint of light grew larger and larger.

'My attitude at this time was quite unlike any of those people about whom the book *Life After Life* was written. I was quite terrified, I did not want to be there, and I was determined that I was not, by God, going to stay.'

As in Kat's hellish experience, Doris's didn't end there, but became peaceful:

'There were beings around me, and they acknowledged my presence. The beings were quite amused at me; there was a "feeling" of laughter within the group who met me. There was one person (if that's what they were) who appeared to be in charge. This one began communicating with me like a firm but loving father would, insisting that I pay attention to him. Slowly, my ruffled feathers became smoothed, and I felt peaceful and calm. I was helped to know that there was nothing to fear in this place. When my sputtering ceased completely, they were able to convince me that it was perfectly okay to be there for a while, that it would be temporary, and that I could return to the delivery room when it was time. So I began going along with this weird experience, and we began to have a question and answer time.

'I would ask the questions and, instead of receiving a wordy reply, they would show me the answer. In the twenty-two years since that experience, only two things remained in my memory. One is a sure knowledge that dying might be unpleasant, but death itself is not a thing to be feared at all.'

Unlike Brenda and Kat, Doris did not attempt to put a label on the 'place' she had been. Likewise, Stewart had a distressing NDE when he lost control of his car on a snowy evening, slid off the road and down an embankment and into a creek. He lost consciousness when his head hit the windshield. He described leaving his body and watching as the icy water started filling his car:

'I saw the ambulance coming, and I saw the people trying to help me, get me out of the car and into the ambulance. At that time I was no longer in my body. I had left my body. I was probably a hundred feet above the accident, and I felt the warmth and compassion of the people trying to help me. And I also felt the source of all that kindness or whatever, and it was very, very powerful, and I was afraid of it, and so I didn't want to accept it. I just said, "No." I was very uncertain about it and I didn't feel comfortable, and so I rejected it.

'At that moment I left the planet. I could feel myself and see myself going away, way up into the air, then beyond the solar system, beyond the galaxy, and out beyond anything physical. And then as the hours went on with absolutely no sensation, there was no pain, but there was no hot, no cold, no light, no taste, no smell, no sensation whatsoever, none. And I knew I was leaving the Earth and everything else, all of the physical world. At that point it became unbearable, it became horrific, as time goes on when you have no feeling, no sensation, no sense of light. I started to panic and struggle and pray and everything I could think of to struggle to get back, and I communicated with a sister of mine who passed away, and pleaded for help. And at that moment, I went back into my body, which had been moved to the ambulance.'

Again, the majority of experiencers said this unearthly realm could not be described in words, and even when pressed to try to characterize it, half still could not find any words that could capture it. Most of the experiencers in our sample focused on *events* in the NDE, such as entering into the light and interacting with other beings, and on their own feelings and thoughts. Many either didn't pay much attention to the physical appearance of the 'other world' – or reported that the 'other world' didn't have anything that could be described as a physical appearance.

On the day after her thirty-fifth birthday, without any warning, Róisín Fitzpatrick suffered a brain hemorrhage that left her in a life-threatening situation in the intensive care unit. She described for me the near-death experience she had in the ICU:

'I became pure energy and realized that "I" still existed even though I was no longer an individual person in my physical body. Instead, I had merged to become one with a greater, light-filled consciousness.

'There was no beginning or end, no start or finish, no

life or death, no "out there" or "in here." It made absolutely no difference if I was in my body; it was not even relevant because I had become at one with this incredibly potent, highly charged field of energy.

'Surrounded by a hushed silence, I became enveloped by undulating waves of opalescent and crystalline light. Simultaneously, there was a feeling of love and bliss that extended on to infinity. From this place everything was possible because only love, joy, peace, and creative potential were real. My understanding of "reality" was turned 180 degrees when I learned that at our deepest level of consciousness we are energy beings of pure love and light who are temporarily residing in physical bodies.'

Margot Grey also described for me feeling bliss when she was stricken by an unknown illness with high fever while traveling in India at age fifty-one:

'A sense of exultation was accompanied by a feeling of being very close to the "source" of life and love, which seemed to be one. I felt embraced by such a feeling of bliss that there are no words to describe the feeling. The nearest I can come to it in human terms is to recall the rapture of being "in love," the emotion one feels when one's firstborn is put into one's arms for the first time, the transcendence of spirit that can sometimes occur when one is at a concert of classical music, the peace and grandeur of mountains, forests, and lakes, or other beauties of nature that can move one to tears of joy. Put all these together and magnify a thousand times, and you will get a glimpse of the "state of being" one is in when the restriction to one's "true heritage" is partially removed.'

Both Róisín and Margot described feelings and events that they experienced in their NDEs, but neither described anything that could be called a 'place.' Because half of the experiencers in my research could not describe a 'place' they had gone in their NDEs, and there was little consistency in the descriptions of the other half who *did* describe a 'place,' none of these images can be called 'typical' of NDEs.

So where does the mind go after death? Do we go to heaven, to hell, or what? Science can tell us what experiencers say about what happens after death, and about the consistency of their reports across different individuals and different cultures. But science at this point usually can't tell us anything about the accuracy of what they say.

I say 'usually' because in some cases, we *can* investigate what experiencers say about the afterlife – if what they say is related to things we can observe in *this* life. Some experiencers' reports may be accurate descriptions of things that really happened, and some may be things they only imagined. But other reports fall into a middle ground of misinterpretations of things that really happened. Jeff crashed his motorcycle during a race and was pinned under his bike. As he lay trapped on the ground, gasoline leaked into his helmet, and he breathed in the toxic fumes before he was pulled out. He was brought to the emergency room with broken bones and abrasions, and quite intoxicated by the gas fumes, which made him frightened, confused, and combative.

When I interviewed him the next day, he was calm but still somewhat groggy. He told me that he had blacked out after the crash, and then awoke in a foul-smelling place where it seemed to him that he was being tortured by beings with eyes, but no other facial features. Some were holding him down and strapping him to a table. Others were inserting needles into his body. Was this a hellish NDE? His memories were somewhat hazy, without the crystal clarity of typical NDE memories. And although the only way he could make sense of his vague memories was to imagine he'd been tormented either by demons or by aliens, he did not claim to *know* what had happened, as most experiencers do after an NDE. Was it instead a toxic hallucination due to the gasoline fumes?

After I talked with the emergency room staff who had treated him the day before, it became obvious to me that this was neither an NDE nor a hallucination. It was Jeff's confused perception of what really happened. The gas fumes had made him so combative in the emergency room that the staff could not examine him, draw blood, or start an IV to medicate him. The medical team, who were wearing surgical masks covering their faces below their eyes, gave him a foul-smelling gas to breathe in order to sedate him. Then, once he stopped fighting, they strapped down his wrists and ankles so they could insert an IV and draw blood. Jeff wasn't having an NDE, but he wasn't hallucinating, either. He was seeing, hearing, and feeling what the team was really doing to him, but in his confused state was unable to understand it.

When Jeff's thinking finally cleared a day later, I was

able to explain to him what I'd learned and help him make sense of his terrifying vision. He was relieved to know that he was neither being dragged to hell nor going crazy, but only temporarily confused by the toxic fumes he'd inhaled. And I was gratified that my taking his story seriously enough to pursue it helped him come to terms with his frightening experience.

So we can't always take reports of an afterlife environment at face value, but we do need to take seriously those accounts that are consistent across different cultural beliefs and personal expectations. People who report near-death experiences need to be listened to. They need space and time to process their mental and physical trauma.

Do we go 'somewhere else'? Even asking that question may be misleading, because 'somewhere else' implies a 'place.' The evidence suggests only that after we die some of us find ourselves still conscious – at least for a while. And because that 'somewhere else' often doesn't look like our usual physical environment, experiencers often label it a different 'world,' such as a 'heavenly realm' or a 'spiritual world.' But that label doesn't necessarily imply a different physical location. We sometimes talk about the 'sports world' or 'the world of entertainment' or the 'political world.' We don't mean different physical places, but simply different aspects of our physical world that we usually don't pay attention to. Is it possible that what experiencers call the 'spiritual world' is not in fact a different physical location but rather a different aspect of our familiar world that we normally don't see?

What about God?

More than two-thirds of the experiencers I've studied say that in their NDE they encountered at least one other person. Two-thirds of those say they met a *deceased* person – an experience that offers at least the potential for some verifiable information. But almost 90 percent of them say they encountered some kind of divine or godlike being. That posed a problem for me, because I couldn't think of a way to test the accuracy of any of their accounts. But I felt I had to pursue them anyway, because many experiencers regard this divine encounter as the most meaningful feature of their NDE. I began looking for some consistent patterns in their pooled stories.

Some experiencers identify the divine being in their NDE as the God of their particular religion. Julia, who had been raised Baptist – though she rarely attended church – described meeting Jesus and His Father when she had a heart attack at age fifty-three:

'I saw Jesus first. He had blue eyes and was smiling. He held out His hand to me. But – funny thing – He didn't talk, but I knew what He was saying. He told me His Father wanted to see me. We floated through the most beautiful place I have ever seen. It was so peaceful. We went to a big, white cloud-looking thing. A man with a long, white beard and long, white hair was sitting on a big, white square thing. He told me I could not stay. I had to go back. I was needed here more, but soon I could come be with Him.'

Julia clearly identified these beings as Jesus and God, consistent with her religious upbringing – as did about a third of experiencers who reported meeting an apparently divine being in their NDEs.

Other experiencers identify the being they met as divine, but do not necessarily identify it as the God of their religious tradition. Suzanne Ingram, who regarded herself as a 'lapsed Catholic,' was rushed to the emergency room after a car accident at age twenty-two. She told me about meeting a being she identified as her 'Creator' who was not necessarily the God of her Catholic upbringing:

'Then another experience began. I recall meeting my Creator. Call this Creator what you will: God, Buddha, Krishna, Allah. It does not matter. I will call him God for simplifying this, but do not refer to any particular religion or God.

'God spoke to me. God spoke for a while; then God told me I could stay there now and that my life would be considered a success. It would be a good place to stay, but I'd have to return to earth in yet another lifetime to accomplish what I hadn't completed so far in this lifetime, or I could go back to earth and continue my life. I believe he said I would complete my mission here on this planet and would be able to pass into the world beyond the door. God opened a door just a crack and let me peek into the light that streamed from the door, and at that precise moment I chose to return to earth and continue my life. Nothing was going to keep me from attaining that place in death, and I knew I wouldn't have to return to earth in yet another lifetime. I recall being very determined to achieve my goal. What that accomplishment is, is still unclear to me.

'I decided to return. I remember both God and I were smiling. God was very pleased with my decision. That decision to return to earth, in itself, was another step to my ultimate destiny.'

Suzanne identified the being she met as divine, but unlike Julia did not label that being as the God of her Christian tradition – and in fact used the word 'God' simply to make talking about it easier.

Rachel Walters Stefanini, who had been raised Protestant, later became an 'eclectic pagan' who worshiped nature in private rituals in her home. She described for me meeting in her NDE both a Buddhist and a Celtic deity, two beings from very different spiritual traditions, when she was unconscious as a result of extreme blood loss from a hemorrhaging cervical cancer at age forty-five:

'I found myself held in the lap of the motherly and gentle Kwan Yin. I adored Her totally and felt a peace and safeness that I have never experienced before in my human life. She held me close and stroked my hair, speaking soothing and calming words to me. I still cannot tell you what She said. In my mind I can still see Her mouth moving, but I cannot hear Her words. All I know is that She brought me peace as She ministered to me. I felt no fear at all, only a deep, abiding peace. The Lady was to my right. To my left the ancient and wise Celtic Lord Cernunnos sat. I was delighted to see Him because He is so ancient, but I was shocked, since I have never called on Him. Lord Cernunnos was quiet and sat with eyes closed as if meditating. I can remember watching the great horns coming out of His head and thinking how strong He must be to sit there so still with all that weight on His head.

'I have no idea how long I lay there in the presence of these two loving Beings. I can remember the cool silky touch of the Lady's clothing against my lower arms and hands.

'I awoke from the surgery with the first thought in my head as, "You're fine, Daughter." I knew the voice to be of Kwan Yin. I knew I was going to be okay.'

The presence of two deities from very different cultural traditions is quite surprising. As Rachel herself explained to me, 'Please note that these two deities should technically have not been there together. Pagan elders and resources will insist that a human should never mix gods from differing systems and backgrounds. Other pagans that I have related this to are also shocked and can offer me no explanation other than, "They must have been what you needed." '

This combining of deities from different traditions (and the explanation 'They must have been what you needed') suggests that the images of Kwan Yin and Cernunnos were formed partly in Rachel's own creative imagination as her personal interpretation of what she was experiencing. In fact, Rachel agreed that the divine beings she encountered may have appeared as images taken from her mind so that they would be familiar to her. 'I am not a Christian,' she said, 'so I am sure my NDE was framed in a way I felt comfortable with.'

John Seidel broke his collarbone and seven ribs in a serious motorcycle accident when he was sixty. He awoke in the intensive care unit having trouble breathing, and an X-ray showed that his entire chest cavity was filled with blood and both his lungs had collapsed. He described for me the NDE he had during emergency surgery to drain his chest:

'The next thing I remember was being guided along in a white world. I was taken into a room that did not have ceilings or walls or corners but seemed a confined space. In front of me was a figure in white robes with long dripping sleeves. He had long hair and a long, bushy beard that had streaks of black and gray. He raised his right hand and pointed over my left shoulder. I felt very warm, at peace and comfortable about my situation and reality. I saw the folds of his robe and his beard in great detail. I woke up and told my wife that Gandalf was in charge. The figure that I envisioned looked like Gandalf in *Lord of the Rings*.'

John had grown up in a family that moved around quite a bit and attended whichever church had congregants his father knew in his new job. He had been to many religious services of different faiths but at the time of his NDE hadn't been to a church service, worshiped, or prayed for many years. When he encountered a white-robed, benign authority figure in his NDE, the name he came up with was not a deity but a fictional wizard from J. R. R. Tolkien's popular fantasy epic. Among all the experiencers in my research who described meeting a divine or godlike being in their NDEs, one-third identified the being as an entity consistent with their religious beliefs, while double that number – two-thirds – said they could not identify the godlike being.

Although many experiencers settle on the label of 'God' or some other familiar deity for the divine presence they sense, some acknowledge that the label is inadequate. Eben Alexander described an all-loving deity, for whom 'the term "God" seemed too puny a word,' and Kim Clark Sharp, who had an NDE when she collapsed without a pulse on the sidewalk at age seventeen, said that 'even the word "God" seemed too small to describe the magnificence for that presence.' Many others simply describe the presence without trying to identify it.

Tracy, a twenty-seven-year-old agnostic who had an NDE when she skidded on black ice and crashed into a tow truck, described to me her sense of merging with a divine presence:

'I felt completely surrounded and taken up in an inde-scribably warm and loving Omnipresence of Light. The serenity and unconditional love emanating from it through me is beyond verbal description. Direct, unimpeded trans-ference of thought, more like a shared knowingness, was washing through every cell of my being. IT was me and IT was not me. I was IT and I was not IT. I was *in* IT, *of* IT, yet still simultaneously my individual unique beingness. I knew myself to be preciously priceless to this Presence of Light and Sound, as if I was an atom of IT. A drop of the ocean is the essence of the ocean, though not the ocean; the ocean is not complete except for the existence of the presence of every single drop of which it is composed. That is how I related to the Light and Sound in which I was immersed.

'I did not see this Presence of Light and Sound so much as I simply, totally knew and loved IT, within and about me, as IT knew and loved me. There was no space, no time, no separation, no duality of anything, as every cell of my being was flooded through and through with knowingness of how all that is, just is; of how it all makes Divine sense, of how all is in Divine Order. Of how loving one another is loving self and loving the Divine, of whom each of us is an atom.

'Like a hand is part of the human body . . . though it is not the body, and though the body is not a complete body without the hand . . . I knew in that moment, and for all time, that I was a unique atom-aspect of this wondrous being. In a quickening of awareness I felt illuminated with an understanding of how each individual is an aspect of The Source. Words do not do the experience justice, any more than they could to what the yellow-pink-gold of this morning's sunrise looked like, or felt like, to one who has never seen a sunrise.'

And Rudy, who had an NDE at age twenty-six when he rolled his car and sustained a massive brain injury and multiple fractures, also described merging with a divine presence:

'I found myself in a soft, velvety, pure, infinite darkness. I was more aware of this infinite vast darkness than I am aware of what I am telling you right now. There is this sense of completeness. I felt together and complete, but confused in my thought.

'Then it appeared: a pin-dot of twinkling, white Light. We became aware of each other, a sense of loving, peaceful oneness. No sense of time had gone by since the awareness of the darkness, as well as throughout the experience. From here is where the experience becomes even more profound and difficult to share: the only way to understand unconditional *Love* is to experience it.

'As soon as the Light became present, I felt as if I was merging – a sense of communing with what I believe is the Light of Love. It seems as though I was moving, traveling, being drawn to, propelled toward this Light. I know I travelled through eternity at incomprehensible speed, yet "being" still simultaneously. As I came closer to the Light, it became brighter, pure white. The Light of Love was the totality of all the richness of all the good qualities, which I have learned became more and more profound in the experience. Peace, tranquility, harmony, oneness, well-being, unconditional Love, and acceptance is everything I had thought God is and so much more. The Light became my wholeness, with a brightness of a quality and richness that I will only be able to understand in empty forms of definitions and physical metaphors until the day I experience it once more; beauty that takes the breath away just thinking about the experience. I went into the Light; I became one with the Light.'

Many experiencers report realizing in their NDEs that we are *all* divine. Anita Moorjani, whose body shut down racked with lymphoma tumors, told me that in her NDE she perceived that we are all part of the Divine:

'In my experience, I *became* the Source, and there was total clarity . . . Because of the nature of my experience, my sense is that at our core, we're all One. We all come from Unity into separation, and then return to the Whole. I feel that my NDE was a glimpse of that Oneness. I could refer to it as *God*, or *Source*, or *Brahman*, or the *All That Is*, but I think different people have different ideas about what it means. I don't perceive the Divine as a separate entity from myself or anyone else. To me, it's a state of being rather than a separate being . . .

'Once we describe this energy with a different word – such as *Source*, *God*, *Krishna*, *Buddha*, or whatever – it can be difficult for some of us to see beyond the name. These terms mean different things to different people, and also seem to impose form upon the infinite. There are often certain expectations attached to these labels, and many of them keep us locked in duality so that we view this energy as an entity separate from us. But Universal energy, like our pure state of consciousness, needs to remain limitless and formless so that it can become one with us.'

Anita is perhaps different from most experiencers because of her mixed cultural background. Raised by Hindu parents in Singapore, where the dominant religions are Buddhism, Islam, and Hinduism; then growing up in Hong Kong, where the dominant religions are Buddhism, Taoism, and Confucianism; and attending Catholic schools to get a better education, she had been exposed to a wide variety of religious metaphors for the divine. On the other end of the continuum, some experiencers were raised without any religious training and came to their NDEs as atheists, only to have their beliefs challenged by what they experienced.

For example, Janice Blouse, whose heart stopped at age twenty-eight when she vomited large amounts of blood from multiple stomach ulcers, told me, 'I was always a professed atheist, but after my experience I know there is God. He was waiting at the end of the tunnel, and somehow I know this. I felt a peace and tranquility I had never known. I find it very reassuring now, because I know our spirit does outlive our body, and that dying is a very pleasant experience.'

And Marcia, hospitalized for two weeks at age thirty-nine with sepsis and not expected to live, described for me traveling in some other realm without her body:

'I found myself traveling in a luminous light. I didn't feel like a person, although I could think profoundly. I felt totally at peace and at home. I moved in an upwards motion, at an angle. I didn't think it then, but now I can describe my movement like being in a hot air balloon, traveling in the air with no sound.

'I could see the white, heavy, flapping robes of Jesus Christ and I knew that was my destination. I don't believe in Jesus Christ, so I remember being confused. I think my confusion stopped me from continuing, and stopped the feeling of complete peace. The peace was still intense, and I didn't want to give up that feeling, but I came back because it was too confusing.

'While I was still recuperating, I could think back on my experience and feel the total peace throughout my body and soul. I had grown up with a Catholic father and Methodist mother, and was trained in both religions. When I was quite young, I made the decision that I didn't believe in Christ or the Trinity. Since my near-death experience, I tried once to go to a Methodist church, but cried through the whole service. I couldn't understand why I was being led up to Christ, because I don't believe in Him. I don't know why I had this experience, but since I can still feel the tremendous peace, I'm glad it happened.'

Whether they think of their spiritual beliefs in terms of a particular religious faith or in terms of a nonspecific sense of connection to the universe, most experiencers say that since their NDEs they've been aware of the presence in their lives of *something* sacred or divine. Among the experiencers in my research, more than four out of five described having a stronger belief in a higher power and an inner sense of divine presence.

Tanya, who stopped breathing when she bled out after a hysterectomy when she was forty-one, described for me this continued sense of the divine:

'This near-death experience never left me, and made me totally aware of the afterlife, the spirit world, and what I believe everyone is really in search of, namely, "Is there really a God?" I had to come to the edge of death to have this answered for me. I know now there is a God and He deals with each one of us personally. If this is what I had to go through to come to the faith I have, I thank Him. I can never be the same person I was before this experience.'

Veronica, who had an NDE when her surgical wound became badly infected at age forty-eight, told me a similar story:

'That experience changed my life. It made me more aware that there is a God. Also, since I was delivered, life has a special meaning now. I no longer take things for granted. Also God has become my best friend. I depend on Him and I seek His advice on everything. I pray constantly and thank God for his goodness to me. Each hour, minute, second is so valuable now, and I try my best to help people. I know I was raised from the dead, and for this I am forever grateful.'

*

And Darcy, hospitalized with serum hepatitis at age twenty-eight, described her ongoing relationship with the divine since her NDE:

'While in a coma in the hospital, I experienced a journey of free movement, with no effort on my part. I don't believe I was floating, as I have the sense I was upright. I was drawn down toward an intense illumination. I heard music and had an intense feeling of peace and tranquility. To me, it seemed I was overpowered by an extreme feeling of acceptance and love. There was no thought of worldly things. I had two young children and a husband, but it was as if they didn't exist wherever I was.

'I saw two images; one I think might have been God, the other Jesus. Between them I felt so loved and contented. They had a conversation and it was determined that I should return the way I had come, as there was purpose for me in this other place. So I went backward, still facing what was before me.

'How my life changed is dramatic. I came from a pagan background, but since this near-death experience, I continue to have close personal encounters with the spiritual world. Thoughts and words proceed from my mouth that I have no control of, as if someone else is speaking through me. I have heard God speak out loud to me on more than one occasion – if not God or Jesus, then whoever I heard in conversation in my NDE. Usually I am being given advice or direction.'

Like Marcia, Darcy's NDE included images of divine beings – God and Jesus – that were familiar to her even though she was not a Christian at the time. Even today, she hedges her bets and refers to her divine advisor as 'if not God or Jesus, then whoever I heard in conversation in my NDE.'

I'm a scientist comfortable dealing with this-world evidence, but I'm out of my element dealing with religious doctrines. And having been raised in a scientific household without a strong sense of the divine, I was uncomfortable with the overwhelming numbers of experiencers who described meeting some kind of godlike being – not just because it was not part of my personal background, but also because it seemed like something that couldn't be verified scientifically. But scientists can't pick and choose which evidence is worth pursuing and which can be ignored. If we claim to be skeptics, we can't reject the observations that contradict our worldview and accept those that agree with our views, without looking at the data. As Sigmund Freud warned us, 'If one regards oneself as a sceptic, it is a good plan to have occasional doubts about one's scepticism.'

Just as with experiencers' descriptions of the afterlife environment, science can tell us what near-death experiences say about God, and about the consistency of their reports across individuals, but science at this point can't tell us anything about the *accuracy* of what they say. And as with experiencers' descriptions of the afterlife, it's difficult for me to know whether these descriptions of God are a reflection of cultural projections. But my ingrained skepticism holds me back from taking these descriptions literally. I'm not saying that these seemingly divine beings are unreal. But experiencers attempting to talk about their encounter with the divine in their NDEs use a variety of labels, be it God, Buddha, Brahman, Krishna, Allah, Source, All That Is – or Kwan Yin or Cernunnos. And many of the experiencers themselves – such as Suzanne, Rachel, and Anita – acknowledge that these labels should not necessarily be taken literally, but represent their brain's attempt to make sense of something they experienced that was beyond words.

Some experiencers recognize the divine beings they meet and are not at all surprised to encounter them, like Julia. Others recognize these beings but are quite surprised to encounter them, like Rachel, Janice, and Marcia. Still others don't feel any need to identify or label the divine being they meet, like Suzanne and Anita. The important point seems to be not how experiencers identify or label the divine beings, but how they *feel* in the presence of the divine. Regardless of the label or the surprise, they consistently describe feeling peaceful, calm, tranquil, 'at home,' grateful, and, most of all, loved.

The other consistency in almost all these NDE accounts is that they generally view the divinity they encountered – even if they experienced themselves as being part of it – as something far greater than themselves. That is, although they may now think of themselves as divine, they recognize that they are just one small part of a far greater divinity. Many experiencers use the analogy of a wave in the ocean to describe this condition. The wave is one small part of the vast ocean and is composed of the same water as the rest of the ocean, yet it maintains its integrity as a distinct wave with its own properties – at least for a while. As Tracy, who had an NDE when she skidded into a tow truck, put it, 'A drop of the ocean is the essence of the ocean, though not the ocean; the ocean is not complete except for the existence of the presence of every single drop of which it is composed.'

I had to accept, at this point at least, that questions about the nature and identity of the divine beings encountered in near-death experiences are not going to be settled by science. But whatever kind of divinity people encounter in NDEs, and however they interpret it, it seemed to be one of the most profound aspects of the near-death experience. And the experiencers' reaction to meeting a divine being, and its ongoing impact on their lives, led me to a bigger question, one that *could* be studied scientifically. That question was: What do experiencers do after going through an NDE? What difference did it make if a person had an NDE or not? And that turned out to be – for a psychiatrist, at least – the most important question of all.

This Changes Everything

John Migliaccio was scuba diving off the New Jersey shore on a windy July day while home from college. The water was rough and the visibility was so bad he couldn't see three feet in front of him under the water. After about a half hour, he started finding it hard to breathe, a sign that his air tank was getting empty. He was about a hundred yards out from the shore, and with the rough waves was taking in a lot of salt water. His throat was starting to burn, and he began to get dizzy because he was hyperventilating.

At that point, things got a bit hazy for John. He remembers feeling afraid that he was too exhausted to swim anymore, and then suddenly he was high above the ocean, looking down on a black body in the water:

'I felt absolute peace and tranquility. I had nothing to worry about. Everything was going to be taken care of. I remember feeling at that point like it was all over. And I felt very peaceful. I felt like I could rest, I didn't have to swim anymore. It was like being in a pool, just floating in a pool. I was aware of beginning to float out with the wave, and then not remembering anything after that. The last physical sensation I felt was floating back out with the waves and then I don't have any recall of anything after that physically. I just remember I felt peaceful. It was a feeling like I was surrendering. It was a relief. It was like I was letting go.

'Two other divers were on the beach. They dragged me out of the water, but I wasn't breathing. They opened up the jacket of my wetsuit and they couldn't find a heartbeat. One guy started giving me mouth-to-mouth resuscitation, and another guy was on his knees pushing my heart.

'I never thought about death before this. I was only seventeen years old. What did I know? But then you have that experience and don't feel afraid to die, if that is what dying is, if that's what my experience of dying would be. Because it wasn't bad; it was nice. It was peaceful. I felt like I would be carried along without having to do anything, without having to worry about anything. I just had that feeling of darkness. It was comfortable; it was serenity. My life didn't pass in front of me. I didn't go to heaven; I didn't go to hell; I didn't go to limbo. I didn't go anyplace. I call it being at rest. It's like a flower going very slowly down a stream in the springtime through meadows. It's the only way I can explain it. And it was sunny and bright and it was peaceful and there were birds chirping, and I said, "This isn't so bad, if that's what it's like. It's not so bad, you know."

'There were two immediate effects of this experience. First, I understood why I was still alive. Second, I no longer feared dying. I was not distressed as other members of my family when my grandfather died recently, and I think my consciousness will persist after death.'

My assessments of what near-death experiences mean for how the mind relates to the brain and for what ultimately happens after death are based on decades of research, but they are only my opinions of what the evidence shows. Although I think I've got pretty good evidence to support my assessments, I know that some people may interpret that evidence differently and that new evidence may show I'm wrong. But there is one thing about which I am certain, about which the evidence is overwhelming – and that is the effect of NDEs on people's attitudes, beliefs, and values. If you take only one thing from this book, I would want you to appreciate the transformative power of these experiences to change people's lives.

When I've asked people who have had near-death experiences how their NDEs affected them, the first answer is almost always – as John said – that it changed their attitude toward death. My research and that of others has found markedly lower anxiety about death in people like John than in people who came close to death but without NDEs. The experiencers tend to have less fear of death and dying and are less likely to avoid the topic of death. On the contrary, they often talk about death as a gateway to another kind of life. Among all the experiencers who have participated in my research, 86 percent said they had less fear of death since their NDEs. Even experiencers like John who don't report visiting heaven or seeing God in their NDEs endorse the belief that there is no reason to fear death when it comes.

Sarah also described finding comfort in death after she hemorrhaged at age twenty-three during childbirth:

'My experience will be with me always. I was not *near* death – I was *dead*, clinically dead, with medical evidence to prove the fact. Since then, dying has often been a source of comfort to me. I've learned to modify my lifestyle as I accepted my chronic illness. But I never, even during my worst time, have feared dying. This lack of fear, I feel, has enhanced a hundredfold my enjoyment of living.

'When I was later diagnosed with cancer, during my surgery, and after, I never forgot what it felt like to be dead. My death did not hurt me, but greatly enhanced my life. Knowing that I will be protected and welcomed, that dying is beautiful and completely peaceful, I have no fear. The warmth, the pull, the welcome embrace of those arms in the tunnel are with me always.

'For me, there was no transition. I didn't see myself leave my body and float upwards. I was just there, in the tunnel, at the end of the tunnel. Dying was beautiful, peaceful, and graceful. I have been dead. I know the truth. And I am not scared.'

*

And George told me that he uses the NDE that he had at age forty-nine when his heart stopped to comfort others facing dying:

'All I can say is that it left me with no fear and a great feeling of peace. If this is death, then I would have to say, "Why fear it?" I saw no holy figure on the "other side," nor did I see any relatives reaching out to me, as other persons have reported. But I did not want to come back. While I am not religious, nor do I believe in heaven or hell, I now think that somehow we make a transition to some other state. Whatever that transitional state may be, it was such a pleasant feeling that I almost have a wish to return to it. In any event, dying does not possess the same fear for me.

'The experience has changed my life in several ways. As a director of social services for a hospital, I felt that the near-death experience has made it possible for me to deal with the fears of dying patients. For now, I'm willing to enjoy life while life can still be enjoyable. But I don't think a week goes by when I don't think, "What is beyond the bright light?"'

Psychologist Marieta Pehlivanova and I tried to identify which particular features of near-death experiences are associated with changes in death attitudes. In a sample of more than four hundred people who had NDEs, meeting some kind of divine or godlike being was associated with increased acceptance of death and with decreased fear of death and decreased death anxiety. Meeting deceased loved ones, seeing a brilliant light, and feelings of joy in the NDE were also associated with acceptance of death. And a feeling of being one with the universe was also associated with decreased death anxiety. To my surprise, a sense of leaving the body was not significantly associated with death attitudes. I had expected that an experience of being free from the body would reduce people's fear of death – and other researchers had guessed this also – but that did not seem to be the case.

As a psychiatrist, this comfort with the idea of death after an NDE made me wonder about people who try to kill themselves and have an NDE. It seemed to me that losing a fear of death might make suicidal people even more likely to kill themselves. But that didn't happen with Joel, whom I met in his hospital room the day after he had attempted suicide. Joel was in a lot of physical pain and desperately wanted out of his pain-racked life, but he'd been afraid he would be condemned to hell if he killed himself. Eventually, though, his pain had become so intolerable that he overdosed anyway, and then, to his surprise, had a peaceful NDE. When he told me about it the following day, I asked him how that experience made him feel.

'My ideas about death are totally different,' he said, shaking his head as he lay in his hospital bed. 'Death was utter bliss. I can't begin to describe it. But I can tell you this, it's definitely something I look forward to.'

'Tell me about that,' I said.

'I used to fear death,' he went on, 'and particularly what my fate would be for all eternity if I killed myself.' He paused, then continued, 'But I did try to kill myself, and it wasn't at all what I was expecting. I was told that I'd made a mistake, but that I was loved anyway. I didn't go to hell. I went someplace . . . well, I don't know. I guess you'd have to call it heavenly.'

'So now you see death as something to look forward to, rather than as something to fear,' I summarized.

'You bet,' he said, nodding. 'I can't tell you everything that happened, but I can tell you this: I can't wait to go back.'

'So what are you thinking now about suicide?' I asked.

'Oh, God, no!' he said emphatically. 'I didn't mean that. I would never try that again. My overdose was a mistake, but I was sent back . . . for a do-over.'

'So help me understand this,' I said, gingerly. 'You're back in your body, with pain that the doctors can't seem to control – pain so bad that it made you want to die. What's keeping you from trying to end it all again?'

'It's true that I'm no longer afraid of death,' he said, 'but I'm also no longer afraid of life. Yes, I'm still in a lot of pain, and I don't see a way out of that for now. But I also see that my pain and suffering are given to me for a reason. I see now that there's a meaning to everything that happens, and a purpose for all our problems.'

He stopped and took a drink of water from the cup by his bed. 'I was sent back for a reason. I have a job to do here. The pain is something I need to learn to deal with, not something I need to escape from.' He paused, as if weighing my reaction and whether he should say more. Then he continued, 'I understand now that I'm more than just a collection of molecules. I have a profound connection to everything else in the universe. The problems of this bag of skin are not that important. There's meaning and purpose to my being back here in this body.'

Joel cocked his head and looked at me. 'You're not convinced, are you?' he asked.

I shrugged. 'I'm a psychiatrist. I take suicide very seriously,' I said. 'You tried to kill yourself just last night. You survived – against the odds – and you're still reacting to the shock of what you experienced and the shock of finding yourself back here.' I paused, then went on, 'What you are saying right now is reassuring, but you're still in a very vulnerable situation. Let's keep talking and see how things evolve over time.'

We did continue to talk for a few more days, and then he was discharged from the hospital and continued psychotherapy with a colleague of mine. His pain never went away, but he never tried to kill himself again.

Joel's changed attitudes were not unique. Remember Henry, who shot himself in the head because of unbearable grief and then had an NDE in which he saw his mother, who seemed to be welcoming him into heaven? That vision seemed to ease his grief and, to my surprise, did not make suicide more appealing to Henry. As he put it, 'I don't think about that at all now. I still miss Mama, but I'm happy now that I know where she is.'

My research with patients who had attempted suicide indicated that about one-fourth of them have an NDE in the course of the attempt. Those who do have NDEs are less suicidal after the event than suicide attempters who don't have NDEs. This appears to be paradoxical because NDEs generally lead to more positive attitudes toward death and lessen the fear of dying. Having spent decades running psychiatric emergency services and trying to help people deal with their suicidal thoughts, this discovery stunned me. However, every study on the topic – my own and those by other researchers – has reached the same conclusion: that NDEs decrease thoughts of suicide.

As soon as I recognized this paradoxical effect, I started asking patients who'd had NDEs during the course of a suicide attempt not only whether and how their thoughts about death had changed, but also *why*. They gave me a wide variety of explanations, but I did find some common themes in what they told me. They most often report that the experience made them feel they are a part of something greater than themselves. Seen in that light their individual losses and problems seem less important. They now value themselves for who they are rather than for their circumstances. They tend to feel that life is more meaningful, more precious, and more enjoyable than they did before their suicide attempt, and they feel more alive than they did before. Many link their heightened personal value and more meaningful life to their belief that death is not the end, and to their sense of being interconnected with other people.

And it turned out that not only for suicide attempters but for most people who have NDEs, losing fear of death means also losing fear of life – letting go of having to be in control all the time, taking more risks, and enjoying life to the fullest. Over the years, I've heard time and again that losing the fear of death often leads to a richer appreciation of life, despite outward circumstances.

Glen was electrocuted at age thirty-six when the hand-held drill he was using shorted out, as he was standing on a fourteen-foot metal ladder. He also had an NDE that changed his view of death:

'This experience happened in 1973 and is still in my memory like it happened this morning. I'm not afraid of dying, only of not correcting negative parts of my life. I'm awaiting death as liberty and a new life.

'This life is a shadow of the next living experience. My life is much richer and more fun now. I see a kind of humor or feel a laugh coming up in my throat even when most people would be sad or cry at a situation most grave. I know that we must hurt and have defeat to win our liberty. When we learn enough, we will be free of this world.

'The thing I remember the most, which was to me the most important, was the peace and freedom from the physical body. Like a weight carried so long, its removal was a promise of my spiritual future when I will die for real because no one will be there to aid me. That will be okay. I'm no longer afraid of death, only of pain and the aging process and parts wearing out.'

Katie had an NDE when she almost drowned at age thirty, when her closed-deck canoe got caught in an eddy and rolled over and she could not get out of the canoe's spray skirt. She described both losing her fear of death and gaining an enhanced joy at living:

'It seemed that the minute I swallowed all the water, every muscle, fiber, and thought in my body totally relaxed, and it all felt so good. Being *so* relaxed but still so *aware* seemed strange, but I totally accepted it. I wanted to. I felt it must be my time.

'I believe this experience has changed my life. I think back on the experience as a good one, a learning one. I feel that you do not have to be afraid of death, that death can actually be a beautiful experience. It has made me want to enjoy all the little things around me, to live my life to the fullest, to take time every day to stop, look, and listen, and just to take the time to see something, really see something, for the first time.

'I may find myself saying, "Oh, the first tulip of spring is up," but I will also take the time to go out and look at it and feel it, and just enjoy it for a while. I get great pleasure from just looking at things, enjoying them and marveling at how wonderful and complex life is. Everything has taken on a much broader aspect.'

Peggy had an NDE at age forty-five when her heart stopped during a hysterectomy. She also described her loss of her fear of death and her commitment to live each day to the fullest:

'During a hysterectomy my heartbeat started slowing down and subsequently stopped. I also had no pulse. The anesthesiologist heard the monitor alarm indicate that I had flatlined, and he thought the monitor was malfunctioning. He checked everything and realized my heart had stopped beating and I had no pulse. He yelled at the gynecologist to stop the surgery and called a code.

'The second my heart stopped, I opened my eyes and found myself engulfed in brilliant white light. Being scared was the furthest thing from my mind. I have never felt such peace, joy, contentment, unconditional love, and total acceptance in my entire life! Nothing on this earth compares to the love I felt. Even the light seemed to sparkle with gold dust that felt like love. Being there was the most wonderful, peaceful, protected feeling, and my heart was so filled with joy, I thought it would burst. I never wanted to leave this place. There was no concept of time: two seconds could have been two days, for all I knew. I just never wanted it to end.

'It was what I wanted to do more than anything, but something made me hesitate: my family, perhaps, or just that I had unfinished business; I don't know. They tell me the whole incident lasted less than a minute. In that time, I got a little glimpse of the other side and what awaits me. Love is the most beautiful gift that anyone can give or receive. We all need to nurture our relationships and express our love to those we care about. I see how fragile and short life is, so I now try to live each day to the fullest. I look forward to dying and have no fear whatsoever. It will be when I can go "home," where I came from. I know that God is with me always. There is a great peace and joy in my heart that wasn't there before, and I have a new zest for life.'

*

These wide-ranging and long-lasting effects of NDEs on people's lives have been the most surprising and yet the most consistent aspect of these experiences. Over the years I've met many experiencers who did not strike me as radically different from anyone else – and yet they insisted they'd undergone major transformations of their attitudes and beliefs. As a psychiatrist, I knew well how hard it can be to help people make modest changes in their lives, often requiring weeks, months, or years of intensive work. And yet experiencers claimed their NDEs overhauled their lives in a matter of seconds. When I first started studying NDEs, I was skeptical, to put it mildly, that these reports could be accurate, and devoted many years to documenting the range of aftereffects of NDEs.

It soon became obvious that after people see a different reality in their NDEs – or a different way of looking at reality – they are changed forever. They continue to regard the world of the NDE as 'more real' than our everyday physical world. And they neither can nor want to put it behind them and return to the attitudes, values, and behavior they had before the NDE. Among the experiencers I've studied, 90 percent said their attitudes and beliefs changed as a result of their NDEs, and more than half said that the effects of their NDEs continued to increase over time. Two-thirds said they felt better about themselves as a result of their NDEs and had an improved mood; and three-fourths said they were more calm and more likely to help others than before their NDEs.

After an NDE, experiencers sometimes compare their new view of the world to walking on a rainy night so dark that it is hard to see, and then a sudden flash of lightning illuminates the sky and they see the road and the trees and everything else in their immediate environment. As soon as that brief flash of lightning is over, they are again left in the dark. But even though they can no longer see what's around them, they remember what the lightning bolt revealed to them, and they cannot deny that the road and the trees still exist.

Several researchers have found consistent changes in experiencers' perception of self, relationship to others, and attitude toward life. Experiencers return from NDEs with a new or strengthened belief in life after death, a feeling of being loved and valued by some higher power, increased self-esteem, and a new sense of purpose or mission. This new sense of purpose or mission in life is often related to an experience of having been sent back, or having made a choice to return to life, in order to complete some work. Experiencers typically return with a sense that we are all part of something greater.

They seem to have increased compassion and concern for others and a sense of connection to – and desire to serve – other people, which often leads to more altruistic behavior. Experiencers tend to see themselves as integral parts of a benevolent and purposeful universe, in which personal gain, particularly at someone else's expense, is no longer relevant. They also report feeling greater understanding, acceptance, and tolerance for others.

The personal changes associated with near-death experiences go beyond what we see in people who have come close to death but didn't have NDEs. Although many individuals who almost die feel greater appreciation for life than they used to, those who don't have NDEs often become more anxious and depressed, withdraw from social activities, and have post-traumatic stress symptoms. They often respond to a close brush with death by becoming more cautious and less likely to take risks. On the other hand, those who have NDEs show a greater zest for life, have a more intense appreciation for nature and friendship, and live more fully in the moment without concern for the impression they might make.

I have talked with people in their nineties who had the experience as children. They consistently say that the after-effects are as strong now as they were many decades earlier. Psychologist Ken Ring developed the first objective measure of life changes following NDEs, which he called the Life Changes Inventory. I started using that scale in the early 1980s with people who had near-death experiences. Twenty years later, I decided to track down these same experiencers and have them complete the Life Changes Inventory again, to see whether the changes they had reported in the 1980s were still as strong. What I found was quite striking.

The most positive changes in attitude following NDEs were a more favorable attitude toward death, toward spirituality, and toward life, and a sense of meaning or purpose. Slightly lower increases, but still significantly improved, were attitudes toward other people and toward oneself. Attitudes toward religion and toward social issues were only slightly improved, and attitudes toward worldly things were more negative than before the NDE. All of these changes were virtually identical after the twenty-year lapse. None were significantly changed since the first time the experiencers were asked.

These observations of profound personality changes after NDEs are not new. In 1865, Sir Benjamin Brodie, senior surgeon to Queen Victoria and President of the Royal Society, wrote about a sailor who was rescued from drowning: 'In one instance, a sailor who had been snatched from the waves, after lying for some time insensible on the deck of the vessel, proclaimed on his recovery that he had been in heaven, and complained bitterly of his being restored to life as a great hardship. The man had been regarded as a worthless fellow; but from the time of the accident having occurred, his moral character was altered, and he became one of the best conducted sailors in the ship.'

I heard accounts of these widespread changes in attitudes, values, and beliefs from hundreds of people after their NDEs, some expressed in very simple terms and some quite eloquently. But I was most touched personally by a letter I received after I'd been interviewed by host Greg Jackson on *The Last Word*, a late-night television talk show in Detroit. It was written by an elderly woman with little education who felt moved to share with me the effects of her NDE:

> Dear Dr Greyson,
>   I really did enjoy your Program on the last Word. I call but the line were all tide up . . . The Lord Bless

*me in 1973 . . . About six o clock in the Morning the
Lord Come in got me in kerry me to Heaven in show
it to me. Lord have mercy, it was the most peaceful
place, I didn't want to Come back I seen my body
laying there waiting for me. I kept asking him to let
me stay he said no I have to go back. You no Dr you
don't think about nothing on Earth. that really a
experience . . . Dr I was so shock. But I wish he
Would have kept me. Oh What a day. It's no fear to
die. I don't think no one want to come back. But Dr
I told my pastor in my husband I think they thought
I were out of my Mind But they can think what they
want. But I no I been there. I no where I'm
going . . . But Dr I were bless, the Lord just bless me
and let me see it. I praise him everyday for that. He
was prepare me for later things in my life. Heaven is
there I no it is. My mouth was open in joy all over
you, a new person, it make you walk right it make
you talk right it make you treat everybody right. It is
something different once you been there. You have a
different out look on life. Material things don't
count. Every body just look good to you. You have a
reason for every man fault. He will make you more
close to him. Dr Greyson I just had to write to you to
tell you it is real in I am now no make believe. I
won't for get it, it's no something I'll ever want to for
get. I don't no what you think, but God no it's true I
was in his hand.*

Katherine

As Katherine's letter suggests, for many experiencers changes in attitudes after their NDEs were just the tip of the iceberg. Below the surface was a deeper and more intimate awareness of meaning and purpose in life, and a sense of connection with something greater. I found these aftereffects extremely compelling, but struggled with what they meant.

What Does It All Mean?

Christine, a successful businesswoman, had an NDE when her heart stopped for ten minutes due to a heart attack at age thirty-seven. She was raised in a Christian family who went to church 'on and off.' As she raised her own children, Christianity played a small part in their lives, and she took them to church 'when it was convenient.' She became a workaholic and a chronic worrier. She described herself as an aggressive person and developed high blood pressure from years of work and worry. She described how her near-death experience changed all that:

'I started having chest pains. Within an hour, I was in the hospital, and the doctor told my husband I was having a heart attack. During the night, my heart stopped seven times. The last time was at least ten minutes. All the words in our language are inadequate to describe the experience.

As I try to describe what happened, it gives me a feeling of reaching and grasping for words that just are not available.

'I had the feeling of being drawn into a million or more lights, the most beautiful, brilliant, sparkling lights. This was not a scary feeling, but natural. My life passed before me from childhood, as if thumbing a book quickly. I seemed to know and understand all things: the why, when, where. Then I saw this large light, like a spotlight with a Being in it, as if this Being was the light. My feelings in the presence of this Being, which was God, were joyful, comforting, peaceful, and lovely. It was as if I had arrived where I belonged: a feeling of totalness and happiness that's indescribable. I wanted to stay with Him more than I have ever wanted anything in my life. It was total contentment. I was reaching for Him and He had his hand out to me. Just before He took my hand, He said, "Your children need you." When He said that, I instantly wanted what He wanted, and it seemed He wanted me to come back. So I didn't take His hand, only because He didn't want me to.

'This experience changed my life. I have a deep feeling of unconditional love and understanding of almost everyone. Things that used to make me upset or mad do not affect me anymore. I worry about nothing. I still have concern for things, but not worry. The thing that is most important in my life is not doing what people think I should do, but what I think God wants me to do. My job is to let people know there is another life, and to care for my family.

'Before my NDE, I was very aggressive, wanting all the

material things life had to offer. I had too much pride, a very active temper, a biased attitude, and a desire to control events that I could. I would never take no for an answer. If I got a negative answer, it was like waving a red flag, and I would charge ahead to fight. I was also a perfectionist, so people didn't like working for me.

'Social work had no place in my life. I really didn't like to be bothered with people at all. I had no time for anyone. As I look back I think money and doing what I wanted to do was most important to me. Social standing was very important to me. I drank too much, smoked too much, and enjoyed it all. Or at least I thought I was enjoying it.

'Since my NDE, the biggest change in my life is my unconditional love and compassion for all mankind. I hurt inside for people who are underprivileged, sick, hungry, homeless, elderly, and needy; also for people who are so unhappy. I feel a deep desire and need to help these people. I give at least one-tenth of all income to charitable organizations. The business does not hold me captive anymore; money is not my master. It's for needs we have and to help others. My temper and pride are under control. Nothing seems to upset me. I have some problems, but ride above them as in a boat above the waves. I have peace, joy, and contentment, but look forward with anticipation to my next life. Each new day is truly that new.'

Many experiencers, like Christine, report that the most meaningful change after an NDE is an increase in their sense of spirituality. What they mean by the term 'spirituality' is the aspect of their personal lives that includes something beyond the usual senses, and a personal search for inspiration, meaning, and purpose, a quest to connect with something greater than themselves. For many experiencers, this includes a conviction that loving and caring for others is of primary importance.

Having been raised without a strong religious faith, I had difficulty relating to phrases that Christine and many other experiencers use, like a 'higher power' and 'what God wants me to do.' But the concepts of feeling connected to something greater than myself and of finding meaning and purpose in loving and caring for others sounded like the values my family had instilled in me as I was growing up. Are experiencers who say their NDEs left them feeling more spiritual using different words for the same values and drives that I have, or are they talking about something else entirely? I found their personal stories compelling, but I wanted some objective measure of their spirituality to understand it more fully.

I wanted to compare people who'd had near-death experiences with people who'd come close to death but *didn't* have NDEs – because nearly dying by itself is a huge event that could well bring about life changes. I used well-accepted, standardized questionnaires that measured various aspects of spirituality, like satisfaction with life, connection to something greater, and a sense of purpose. What I found was that those who had NDEs were significantly more satisfied with life, more open to positive new directions in life, had more positive changes in relationships with other people, felt more personal strength, had greater appreciation for life, and felt they had undergone greater spiritual growth as a result of their NDEs. In addition, many people reported that since their NDEs they'd felt driven to engage in a quest for further spiritual growth.

Elizabeth had an NDE at age twenty-eight when one of her Fallopian tubes ruptured, ending an ectopic pregnancy. She described her subsequent quest for spiritual knowledge:

'I had stopped formal education after high school, and was not interested in religious, philosophical, or scientific matters. However, after my NDE, I embarked on a lifelong quest for knowledge in just those areas, and this was years before the printed experiences of others could have influenced my behavior.

'An almost insatiable thirst for knowledge in the subjects of science, philosophy, theology, and what is called metaphysics has dominated my life since the NDE. I have experienced several of what I call "mild" cosmic consciousness episodes and have access to a spiritual library through dreams, helpers, and meditation practiced for thirty-five years. I also have a strong sense that everything in the universe is connected.

'I feel the most important things are seeking and sharing knowledge, and receiving and returning love. I feel strongly it is the spiritual that is important, and that the dogma and doctrine of organized religions are man-made, and for that reason, subject to flaw, and as history has shown, not too effective. I don't necessarily follow church teachings but rather the guidance of the inner spirit.

'The thirst for knowledge is a daily drive with hours spent researching a myriad of subjects. Learning and knowledge are things you can take with you when you pass through. The secret's in the seeking, and the pursuit goes on eternally. That each is responsible for one's own actions and beliefs and progress toward spiritual enlightenment is my life recipe.'

Psychiatrist Surbhi Khanna and I found that people who had NDEs described a greater sense of well-being from their new spiritual attitudes and strivings, which helped them cope with challenges. They also reported more daily spiritual experiences, such as feelings of awe, gratitude, mercy, compassionate love, and inner peace than did people who had come close to death but *didn't* have NDEs. Our studies and the research of others also found that people who have had NDEs report a heightened sense of purpose, increased empathy, awareness of the intercon-nectedness of all people, and a belief that all religions share certain core values. NDEs often lead to a paradoxical *decrease* in devotion to any one religious tradition, despite a greater awareness of guidance by and connection to a higher power.

Spirituality *can* be connected to religious traditions, but many experiencers describe their spirituality as an inner feeling that is independent of any religious practices or beliefs. They often report that they feel such a strong per-sonal connection to the divine that religious observances seem unnecessary. Many experiencers describe adopting a form of nondenominational spirituality since their NDEs, in which all religious traditions are valued but no one religion is given precedence. Katherine Glenn had an NDE from respiratory infections she contracted in the hospital at age twenty-seven, while recuperating from surgery. She told me she saw in her experience that the core of all religions is essentially the same:

'The enormity of this event had such an impact on my life, it opened windows and doors to me that I never knew existed. I was shown that religions are like jars of jelly on a shelf, only each jar had a different label put there by man. It is all jelly, it's all sweet.

'There are many paths up the mountain to reach God and it really doesn't matter which one you take, because when you get there to that mountaintop it is all the same love, light, peace, harmony, gratitude, wisdom, truth, and victory for everybody. There are no religions in heaven, just "jelly."'

I found this spiritual growth most striking in people who were staunchly materialistic before their NDEs. Doug was an atheist who had always scoffed at the idea of a spiritual realm. Then at age seventy-one, he had an NDE when he bled internally from a ruptured artery going to his stomach. He described to me his reluctant acknowledgment that spiritual experiences do indeed happen:

'At about two a.m., I awoke with stomach pains. I was unable to vomit or have a bowel movement and passed out. My wife heard my groans and called an ambulance. In the emergency room, the doctor called in a surgeon, and after several tests they decided I might have a ruptured spleen and started to operate. By this time I had lost considerable blood. The doctor discovered a ruptured artery going to my stomach and I was very close to death.

'During that operation I had an NDE. In the NDE I was next to a wall, bathed in light. On the other side of this wall there was nothing: just total darkness.

'Then I was given a choice. I do not know who gave me the choice, but for all of my skepticism about spiritual things, I do believe that I truly had a choice. I still believe that. My choice was I could either take the "Express Checkout" and die with no pain. Or I could live and then have the probability of facing lots of suffering, hospitalizations, and then dying anyway. On the other side of the wall there was nothing: just darkness.

'I decided not to take the Express Checkout because I could see nothing on the other side of the wall, where I would have gone, but darkness. I also felt it would be worthwhile to find out if the dismal prospects of suffering and then dying anyway might not be true. If it were, I could decide to die the next time. The choice had its foundations in what I believed about death. I thought when a person dies, that is the absolute end, no different than a dog or bird.

'As I think of it now, I believe the spiritual part of my experience was that I truly had a choice; no one made it for me. I am not a spiritual or religious person. I am a college graduate, and was born and raised as a Catholic but ceased practicing that fifty or sixty years ago. And there is no other religion I would adopt.

'What did I learn? First, to be thankful that I had a choice. And second, to live each day as well as I can.

'I really believe I was given a choice. Now that is spiritual, isn't it? And I do *not* believe in such things. Yes, I do wonder about the spiritual part of the NDE. I do not believe in spiritual things, but one seems to have happened to me.'

Likewise, Naomi, a pediatrician who had always considered herself an atheist, described becoming more compassionate and less competitive after her NDE at age thirty-four when she bled out due to a hemorrhaging stomach ulcer:

'I remember the spring after this event occurred with startling clarity. Everything in the environment took on an almost magical quality, as if I was seeing everything for the first time. Trees and flowers blossoming took on new dimensions that I had not ever appreciated before; I almost felt as if I was on a chemically induced high. I will certainly never take being alive again for granted. I also felt that when faced with death again I would be fearless, as this was clearly not a negative experience. I have used this insight to help the families of many of the handicapped and terminally ill children I take care of, with good results. I also developed a strong spiritual sense and now strongly believe in a higher power, where I had previously been essentially an atheist.

'No other experience to date has had such a profound impact on my life. I am much less striving in the workplace. I also feel that material goods, although nice, do not define the spirit or essence of the individual. My life is more balanced than ever before. I am much more open to meditation and other "alternative" medical techniques. I am now attempting to use lifestyle modifications and not medications to control my medical problems. I feel I have developed more compassion for my patients and have become a better doctor for this. I am still integrating many aspects of this experience, and find it is good to contemplate it from time to time to refocus myself and see the larger picture. I suspect it will always be a source of growth for me.'

These claims of spiritual growth after NDEs are very common. No matter how they are worded, they tend to focus on a sense of connecting with something greater than ourselves and an emphasis on the importance of loving and caring for others. Several studies have shown that experiencers express greater compassion and concern for others than they did before their NDEs, and that they are willing to go out of their way to help other people.

This is essentially what we know as the Golden Rule: 'Do unto others as you would have them do unto you.' Every major religion has some variant of this as one of its basic guidelines. It appears in an ancient Egyptian papyrus from 500 BC, in the writings of the Greek philosophers Sextus and Isocrates, in the Book of Leviticus in the Old Testament, in the books of Matthew and Galatians in the New Testament, in the Babylonian Talmud, in the Muslim Koran, in the Hindu Mahabharata and the Padma Purana, in the Buddhist Udanavarga, in the Jain Sutrakritanga, in the Analects of Confucius, and in the Taoist T'ai-shang Kan-ying P'ien.

Religions that believe in God promote the Golden Rule as a divine command, whereas religions that do *not* believe in God promote it as a reasonable guide to living a fulfilled life. As writer Dinty Moore put it: 'If there is a God, I should live my life according to principles of kindness, compassion, and awareness, and if there is no God, well then I should live my life according to principles of kindness, compassion, and awareness anyway. How wonderfully simple.'

Recent work in neuroscience suggests that the universal nature of the Golden Rule is the result of an unconscious brain mechanism that evolved over millennia to help us survive in groups. The universality of this spiritual precept common to almost every religion is reflected repeatedly in the accounts of NDEs. Experiencers often describe the Golden Rule not as a moral guideline we should strive to follow, but as a description of how the world works, a law of nature as inescapable as gravity. They often say they have experienced this natural law firsthand in their NDEs, as they feel in their life reviews the effects of their actions upon others. Though they don't feel punished or judged for their misdeeds, they do receive back as part of their life review everything they have ever given out, measure for measure.

Tom Sawyer, who had an NDE when the truck he was working under fell and crushed his chest, found that in his life review he experienced all his misdeeds from the perspective of his victims. He described for me reliving a memory of beating up someone with his fists:

'I saw myself at age nineteen, driving my hot rod pickup

truck down Clinton Avenue. A man darted from behind a van and almost ran into my truck. It was summertime, the windows were rolled down, and I inched up toward him. I said to him rather sarcastically, "The next time you really ought to use the crosswalk," whereupon he yelled some four-letter words at me and reached through the window and slapped me across the face with an open hand.

'Well, I pulled the keys out of the ignition, stepped out of the truck, and I beat that man up, hitting him many times. He fell straight backwards, hitting his head on the street. I almost killed that man, but I wasn't thinking about him. I was indignant. The guys from the gas station across the street came running over. I said, "Well, you guys saw him hit me first." And I very methodically got back into my truck and drove away.

'Now it's life-review time! I can follow the adrenaline rush from the center of me outwardly, can feel the tingling sensation in my hands and experience the warmth of my face getting red. I can feel the rage that this jerk had violated my calm pursuit of happiness. I never knew that man either before we had the altercation or after. But in the life review I came to know that he was in a drunken state and that he was in a severe state of bereavement for his deceased wife. In the life review, I saw the stool in the bar where that man had his drinks. I saw the path that he took down the street for a block and a half before he darted from behind that vehicle into the path of my truck.

'I also experienced seeing Tom's fist come directly into my face. And I felt the indignation, the rage, the embarrassment, the frustration, the physical pain. I felt my teeth going through my lower lip. In other words, *I was in that man's body, seeing through that man's eyes.* I experienced everything of that interrelationship between myself and that man that day. You better believe that I was in that man's eyes. And for the first time I saw what an enraged Tom not only looked like but felt like. I experienced the physical pain, the degradation, the embarrassment, the humiliation, and the helplessness in being knocked back like that.

'After I stepped out of the truck, I hit that man thirty-two times. I broke his nose and really made a mess of his face. He went straight back and hit the back of his head on the pavement. Okay, "he hit me first." Try that in your life review! I experienced all of that, right to the man's unconscious state. My life review included experiencing the event from an outsider's point of view, from a third-person viewpoint. This all happened simultaneously, seeing through my eyes and through his. During this life review I watched everything unconditionally. It wasn't judgmental or negative. I had the experience of observing something without any emotion or righteousness, or judgmental attachments.

'I wish that I could tell you how it really felt and what the life review is like, but I'll never be able to do it accurately. Will you be totally devastated by the crap you've brought into other people's lives? Or will you be equally enlightened and uplifted by the love and joy that you have shared in other people's lives? Well, guess what? It pretty much averages itself out. You will be responsible for yourself, judging and reliving what you have done to everything and everybody in very far-reaching ways.'

*

Some people dismiss these spiritual lessons from NDEs as clichéd, well-trod religious platitudes. Well, yes, they are. The reason the lessons of NDEs may sound banal and recycled is that we *have* heard them all before. Time after time, experiencers have told me that their NDEs did not reveal things they hadn't known, but rather reminded them of things they had once known but long ago forgotten or brushed aside.

Kim Clark Sharp had an NDE when she collapsed on the sidewalk at the age of seventeen. A nurse standing nearby looked for but couldn't find Kim's pulse and told someone to call the fire department. The firemen arrived quickly and hooked her up to a portable ventilator machine because she had stopped breathing, and they started pounding her chest. Kim described to me her experience after she collapsed:

'Suddenly, an enormous explosion of light erupted beneath me, rolling out to the farthest limits of my vision. The light gave me knowledge, though I heard no words. This was discourse clearer and easier than the clumsy medium of language. I was learning the answers to the eternal questions of life, questions so old we laugh them off as clichés. I felt as if I was re-remembering things I had once known but somehow forgotten, and it seemed incredible that I had not figured out these things before now.'

Of course, the true test of spiritual growth is not what people feel or say but whether it translates into everyday life. As educator Frank Crane put it, 'The Golden Rule is of

no use to you whatever unless you realize it's your move.'
Fran Sherwood had an NDE during emergency abdominal
surgery when she was forty-seven. She described the
importance of focusing not on the experience itself but on
acting on what she learned:

'All of this had and still has such a profound effect on
my life that I have not been the same. Yet, I am still me,
perhaps a freer person than before. All my values have
changed and are still changing, becoming clearer. There is
often a sense of hunger for a deeper involvement with my
fellow man, and I am always seeking a closer touch with
God. And along with the daily routine of living, I try to
improve whatever I can, wherever I can, and to spread the
message of Love, in all the small ways that we do.

'The experience is valid and there is a certain joy and
awe in relating it. But the moment comes when the experience ceases to be the focal point. You have to really look
upon it as only a beginning – a new birth, if you will – and
from that point on you begin to grow. This time, the
growth is a new reality; it points you to becoming involved
with others. The self begins to dwindle away and though
you may try to hang on to the near and dear of self, you
really have to let go. For if you do not, you will be negating
the purpose you now have. This growth is for your good
and ultimate happiness.

'Over and above the talking about it, the sharing, has to
then come the *action* – not that you have to stop talking or
sharing, but now included in that is the action, the action
of doing what we were sent back for. It may have been presented to each of us in different ways, but the same message

comes out loud and clear. We all know what it is, and though it can be said in a thousand ways, there is only one word that says it all: Love. And the message is this: "Just as I have loved you, you also must love one another." This is an irrevocable Truth.'

And in fact I found the most impressive aftereffects were not the changes in attitudes, but the dramatic changes in lifestyles that often followed NDEs.

A New Life

The shrapnel entered through the armhole of Steve Price's bulletproof vest just as he was firing his rifle, and a mortar fragment pierced his lung. When the medics finally reached him, they airlifted the twenty-four-year-old marine from the Vietnamese jungle to a military hospital in the Philippines for surgery. During the operation, he left his body and had a blissful experience of light, warmth, and peace. The burly, heavily tattooed, self-described schoolyard bully became tearful when he described to me what happened:

'Suddenly I realized I was up near the ceiling, looking down at my body. A brilliant, white light embraced and encompassed me. It took me in. I felt such warmth and peace and was bathed in the most peaceful, joyous feeling imaginable. I found myself in a place like the Garden of Eden. I didn't say the word "God" for a long time, but now

I can say that the light was God. It was like the most loving mother embracing her infant, only a million times more than that. On the other side of a stream bubbling through the garden was my long-dead grandfather. I went to move toward him, but it was over.'

After his recovery, he tried to return to battle, but found that was difficult:

'I led my unit. I did all the things I was supposed to, but I couldn't shoot my gun. All I ever wanted was to be a marine, but I realized I could no longer do my job. The NDE had an incredible effect on my life. No matter how hard I tried, I couldn't fire my rifle. Eventually, I left the Marines and now I work as a lab technician. I joined the National Guard, because it helps people instead of killing them. I'm now mild-mannered, thoughtful, and a lot different from the hard-charging, macho marine I once was. I've become so sensitive that when other people hurt I can feel it. I talk to people about metaphysical things. Before, I would've made fun of people who talked about such things.'

Many experiencers like Steve report that after their NDEs their previous lifestyle no longer felt comfortable to them or was no longer fulfilling. As a result, some do change their professions, as Steve did. Among the experiencers I've studied, one-third changed their occupations as a result of their NDEs, and three-fourths reported a marked change in their lifestyle or activities. These changes are most dramatic with experiencers who had been in a profession that involves violence before the NDE, such as law enforcement or the military. Joe Geraci, the thirty-six-year-old policeman who almost bled to death after surgery, described a life change comparable to Steve's after his NDE:

'I was a no-nonsense, hard-nosed cop. My NDE changed all that. I left the hospital a completely different man. After being a cop used to bloodshed, I found I could not watch TV because it was too violent. After endangering myself and my partner on patrol because I couldn't fire my gun, I quit the police force and retrained as a teacher. Although I found teaching high school fulfilling, I often found myself reprimanded for becoming too involved in my students' personal lives.'

And Mickey, who collected money for the Mafia in the 1970s, also had a profound reversal of his personality and transformation after an NDE. Immersed in the material world of 'quick cash and getting ahead by butting heads,' his last job prior to his NDE was as chief steward of a mob-owned resort, where one of his prime functions was providing sexual and other kinds of illicit entertainment for celebrities who performed at the hotel. In this capacity, he was in charge of a number of high-class prostitutes whom he often treated roughly. Then, during a heart attack, he had a near-death experience that included blissful feelings, seeing the light, and communicating with a divine being and with a much-loved brother who had died many years before.

He came back from his NDE with aftereffects similar to

those of Steve and Joe. He felt that cooperation and love were the most important things, and that competition and material goods were irrelevant. That change in attitude didn't sit well with Mickey's Mafia friends, but they let him leave the family circle. It was his girlfriend who complained when he changed careers and started helping delinquent children and victims of spousal abuse. One day after he was out of the hospital and they were eating lunch, she burst out crying and told him, 'You're not the same person anymore!' When he asked her what she meant, she replied, 'You're not concerned with things of substance anymore,' meaning money and jewelry and fast cars. The relationship soon collapsed. Mickey compared his attitudes and behavior before and after his NDE in graphic terms:

'Before the experience, my attitude was that people have to help themselves. You know, if they don't help themselves, to hell with them. I had a pretty cynical attitude toward people. I couldn't imagine myself as any sort of helping professional before the NDE. But afterwards, I'd find myself counseling people. I'd find myself listening to people. They said, "You really listened to me. You really understand how I feel inside." *Before*, I would say, "Listen, pal, I ain't got the time. God helps those that help themselves. So get your butt out there and help yourself. Because it's war out there, on the street. Make sure you always cover yourself out there, because it's a war."

'Before, I thought, "I have to make my way the best I can. Survive." Whenever I started to feel sorry for somebody, I'd say to myself, "Goddamn it, I'm not my brother's keeper!" I was hard-bitten. But after the NDE, my whole outlook

changed. I can feel when people are in pain. Before, sometimes I had to *cause* people pain. I couldn't do that anymore
after my heart attack. Before, I had to take care of number
one. If I gave myself to a job, in gambling or whatever it
might be, I would carry it out. That was the rules.

'The experience made me more sensitive to and aware
of others' pain. I still get very teary about others who are in
pain. People I know can't understand that. Sometimes I sit
down and look around at myself and say, "What the hell
am I doing here? I could be making ten times this money."
But I don't want that. My needs are simple. I'm very content. I could live in one room. I used to have a big Cadillac,
a luxury apartment. I needed those things. They were
necessary for my identity. Now, to tell the truth, it doesn't
make any difference whether I make ten dollars a day or
ten thousand dollars a day. It doesn't matter; it doesn't
mean anything. That's not what's important in our trip
here on this earth.

'Right now, I'm up to my ears in bills. I got all kinds of
bills. But it doesn't really upset me. I'm not driven for the
money itself anymore. I can't do the kinds of things I used
to anymore to make quick money. I can't do that. It's not
that I now think that God, the Big Thumb in the Sky, is
gonna get me. It's something, instead, between Him and
myself.'

This kind of attitude change that results in a career shift
also occurs in people whose prior profession was based on
competition. Emily had an NDE at age forty-nine when

she almost drowned while swimming in the Gulf of Mexico. She described her loss of a 'killer' instinct and values after that event:

'I am a very successful real estate broker of seventeen years. Since this event, I have now given my business over to one of my sons and have joined the Lutheran Church Home, where I am marketing housing for the elderly. I can honestly say that each day I now live to the fullest and would never, never be afraid of dying. It can only be a superb trip!

'This experience has left me with a heightened compassion and intense feeling that anger and hatred are wrong and wasted emotions. The joy of being alive took away a big part of the feisty attitudes I was known for; the "killer" instinct to do a deal was gone! Making big bucks is no longer a priority. The empathy of helping people hurting is from the heart. I treasure time with my sons as never before. I used to walk out on family get-togethers to "make a deal," but no more! Material things are simply not worth the hassle anymore. And I never miss the chance to tell those I love how much I care; I might not have another chance.'

Likewise, Gordon Allen was an entrepreneur, a ruthless and successful financier, who became acutely ill with congestive heart failure at age forty-five. After his NDE, he severed all links with his business and left the world of money far behind. He became a licensed counselor and used his new understanding to help others change their lives:

'Immediately the thought was communicated to me that all the skills and all the talents and everything that I had been given, which I had been very, very, very blessed with, were for a purpose greater than the purpose that I used them for, the purpose of making money, and that itself wasn't it, and there was another purpose for it, and that they should now be applied in some ways that would be shown to me. Absolutely, that's the moment my life changed.

'When I came back, my heart was filled, and you would describe this as being on fire. Your heart feels like your heart's on fire, and it's on fire with love, okay? The sensation of love that I experienced as I was going through the out-of-body experience has retained itself. I'm there; it's in me. It hasn't gone away, hasn't changed.

'So I'm trying to understand where it's going to go, but I've decided I was not going to try to save anything that I had from my past life in the financial world or the business or anything. So those people who were most vehement about the fact that I wasn't going to work in the real world anymore as financial guru, and that they weren't going to be able to make a lot of money off of me anymore – by whatever method that was – I called them up. And I say, "Hi, Bill," "Hi, Jack," or whoever it was, and "This is Gordy. Do you remember how . . .?" "Oh, yeah, Gordy," and you can hear them pull back off the phone because they're waiting for the hit, 'cause in the old days that might have been a hit coming after them looking for the money or whatever I was doing. And I said, "Hey, you know, the last time we talked, I wasn't really too happy about the way that came out, and I think I would be less than candid if I didn't say I was not being good to you, that I was not being loving to you, and I just wanted to call you and ask your forgiveness for whatever I might have done to you."

'Now, if you want to hear dead air on the telephone, do that. You have total silence, and then there's a little stutter and they say, "Well, yeah, I guess so," or whatever, and that's the end of it.'

Interestingly enough, these claims by experiencers that they no longer feel addicted to worldly things doesn't mean that they give them up. In fact, many of them say that because they no longer feel driven to accumulate possessions, they paradoxically feel freer to enjoy material pleasures more fully. They don't renounce worldly possessions, but they feel less attached to them. They no longer define themselves by what they own.

One day, Tom Sawyer, the highway department supervisor whose truck crushed his chest, arrived for a meeting at my home with an ear-to-ear grin, having driven five hours in his Cadillac Eldorado. True, it was a used car, but owning such an extravagant car didn't seem to fit my image of a blue-collar worker who claimed to have no interest now in material possessions. He didn't see the contradiction. He said he enjoyed the luxurious seats and handling of the Eldorado, far more pleasurable than any car he'd ever driven. It was his enjoyment of all life had to offer, including its pleasures like sweet tastes and smooth rides, and not his desire to own things, that had led him to buy this used Cadillac. And sure enough, when a couple of years later, he could no longer keep up the payments, he didn't blink an eye when it was repossessed. *Owning* the Cadillac was never the point. It was all about the thrill of driving it for a while.

Some people feel their NDE not only changed their priorities, but saved them from a destructive lifestyle. Dan Williams's heart stopped while he was withdrawing from drugs when he was thirty-seven. He described for me a remarkable transformation after his NDE:

'I lost everything to an addiction to prescription and street drugs. I went to drug treatment centers at least nine times, was charged with over a dozen drug charges as well as several alcohol-related charges. I had given up on any attempts to get clean and sober; I had pretty much given up on life.

'How does a guy go from a hopeless drug addict, nearly homeless and broke, stealing drugs to support my habit, to owning senior living communities, often sitting and comforting the dying? I still cannot fathom the amazing transformative power of my NDE. The effects still amaze me and my wife. It was truly a blessing. I was an extreme skeptic who used logic and scientific method to resolve life's issues. I was spiritually ignorant and lost.

'I was arrested again in October 2003 and had been told by substance abuse doctors and counselors that I would probably die from the withdrawal rebound effect from the fifteen years of daily pill abuse, so I prepared to die in jail. On the seventh day in jail, during a grand mal seizure, my heart stopped. I was revived and rushed to the hospital.

'I remember what seemed like I was traveling out of my body and then coming back, like I could leave but had nowhere to go. I felt as if I was floating or suspended in a milky black area. I began to experience a feeling that I was immediately drawn to. This feeling was not of this world: all my pain was gone. My first thought was literally, "There isn't a drug on earth that can make you feel this way." I just can't put into words what this felt like: perhaps nirvana, purest form of love, whatever. I just knew I would never be the same. I felt I was moving toward this feeling through the milky stuff, being drawn to it.

'At this time, I could feel a presence with me. I guess you could call it a guide or angel, but I never did see an actual figure or person. There was nothing but light all around and it didn't hurt my eyes. The brightness of the light was off the charts. The presence was guiding me in the light; it seemed to be beside me or behind me. I trusted the presence. I felt as though I knew this presence. There was a feeling of peace. There were no fearful feelings. It was like perfection. I really can't describe it.

'At one point, I was able to see my addiction for what it was and confront, or I will say fight it. It represented the worst of me and I was pissed. I was ashamed of what I had become. For the first time in my life I felt rage, and for the first time in my life I had the upper hand on my addiction. In my NDE, I fought and won and now the addiction was over. I've never desired nor taken a drug or drink since that happened sixteen years ago. There were more components of the NDE, but this was the most impactful. Instead of taking from my fellow man, I am of service to my fellow man. I know who I am today and where I'm going. I am not lost anymore.'

Of course, people who have NDEs are still people just like the rest of us, complex mixtures of emotions and thoughts, good and bad, pleasant and unpleasant, altruistic and self-ish. Tom Sawyer was a perfect example. His wife, Elaine, talked about the mixed blessing of his NDE. Her dark eyes flashing, she described what Tom was like before his NDE: 'He was violent, throwing shoes and things at me, shouting that he was the head of family and we had to do whatever he said.' Then she smiled and added, 'But after his NDE, he's become compassionate, gentle, and considerate. He hasn't raised a hand to me or the kids once in all the years since his NDE.' On the other hand, she complained that he now treated *everyone* with that same loving compassion, and at times she felt neglected because he was off helping a stranger in need. Shaking her head, she said, sighing, 'He couldn't care less about the fact that our sofa is falling apart and we need a new one. We have the money saved to buy one, but he doesn't see the point.'

And just as NDEs don't turn people into saints, so too life after NDEs is not all rainbows and butterflies. I found that NDEs sometimes result in serious problems.

Hard Landings

Most of the aftereffects that experiencers report after NDEs are positive effects. But how could such a profound experience that differs so radically from everyday life not lead to problems as well? In fact, not all the aftereffects of NDEs *are* positive. Some experiencers have difficulty reconciling their NDEs with their religious beliefs. Some find it hard to resume their old roles and lifestyles, which no longer have the same meaning, or to communicate to others the impact of the NDE. Some experiencers report anger at still being alive – or at being alive *again*.

Cecilia, a sixty-one-year-old teacher, began vomiting so violently one night that she felt an enormous rip in both her sides. The pain was excruciating, her fever rose rapidly, and she could not stop shaking. Her teeth rattled, and no amount of blankets could keep her warm. Her husband drove her to the hospital, where an X-ray showed a fist-size mass on her right side. She underwent surgery, which found a gangrenous appendix that had ruptured, causing a widespread infection in her abdominal cavity. She described a near-death experience as her fever raged:

'I felt myself being transported into what I can only attempt to describe as universal blackness. It was as though I had been transported into outer space and was looking down at what I knew was Earth. It appeared as a globe with a bluish glow to it. It was about the size of a child's large rubber ball from the vast distance I had traveled. I received the message, "It's okay to go. The seeds have been planted; your work will go on."

'I experienced a wonderful feeling of peace and freedom. I saw my students going out and assisting others, and I knew the work I loved would go on without me. I had a sense of how small we are as individuals in this vast universe. Of course, people close to me would mourn my death, but in reality my passing would be as insignificant as one less grain of sand on Earth, or one less drop of water in the ocean. Death is nothing to fear. It brought me an incredible feeling of love and peace. I also saw how insignificant are the material things we leave behind us. I felt ready to go, reached my arms out to two spirits who were in the room watching, waiting – and then they began to back away, leaving me behind! I pleaded, "Here I am; take me with you," as they gradually faded away.'

But Cecilia's blissful experience ended with a return to her pain-racked body:

'My journey back into the real world began. My

recovery proved slow and tedious. My body was healing, but I regretted that I had not died. The magnificent peace that I had experienced was something I could not shake. I went through weeks of depression. Everything became such an effort: dressing myself, tying shoes, chewing food, swallowing, driving a car, turning the wheel, walking up stairs, turning doorknobs, walking, walking – everything, even talking! Carrying around the physical body just seemed like too much effort. I remember thinking I might have to wait another twenty years before I would have another opportunity like that. I knew of course that death must come naturally in order for me to enjoy such peace.

'I did not know how to climb out of this hole. It just seemed to be getting deeper and deeper. I looked everywhere I could in a desperate attempt to find answers. I bought myself a notebook to keep a journal of how I was to get through this. My first entry was written to God in anger. I asked, "Why am I alive?"'

For others, the distress is not so much anger as sadness or regret. Lynn, who was hit head-on by a drunk driver while riding her bike, described her feelings after she awoke from a coma, wanting to return to her NDE:

'I didn't want to be back. I feel overwhelmed when I think about how my experience has changed and is changing my life. After my accident, after I got out of the rehabilitation facility, I was so depressed that all I wanted to do was die. My parents took me to the psychiatric center and I refused to talk to anyone for about three months. All

I could think about was how much I hated it here, and that I wanted to go back to that place. I felt like I had lost everything I had ever worked for, plus my identity. That place was so wonderful, and I wanted to go back so badly. I was so angry, and people on earth just seemed so mean, and nobody could understand. A psychiatric center also isn't exactly the place to try to talk to someone about this stuff.

'All I did after my accident was sleep. I was also in a lot of physical pain, and I was so confused, because I was now aware that I know nothing. I was furious that I had that experience; it's not just some great thing that happens to you. It makes life harder; at least, for me it did. I have always felt like, "Thanks a lot; why couldn't you have told me just a little bit more, or nothing at all?" It really took me some time to adjust emotionally to the fact that when you die, you don't just stop existing. It really got to me; I just couldn't figure things out.

'Around that time, the Gulf War started up, and I just couldn't understand why so much bad goes on in the world. I felt so much pain and I didn't understand any of it, or why it had to happen. That is one thing that was so confusing: if there was this Creator who is so absolutely incredible you can't believe it, why is He letting innocent kids, people, and animals be harmed?'

It took Lynn a few years to feel comfortable being alive again, but eventually she came to appreciate being given a second chance:

'It has been years since my accident, and I feel like I am just starting to relax and learn things. And I know it sounds weird, but I am so excited to be alive. Life is so incredible;

it is such a gift. I'm not scared to die, but I'm just not ready yet. I have too much I want to do first. I was sitting outside, looking at the moon, the other night, and it is so incredible to be able to see another heavenly body from here. I just feel like I have taken so much for granted, especially time. I'm not saying that I don't still have bad days. I have awful days; but I am glad to be alive. That is really different from the way I used to feel right after my NDE.'

For some experiencers, the problem is not anger or depression but crippling confusion. Louise Kopsky had an NDE at age twenty-nine while she was being anesthetized for childbirth. She heard her heartbeat getting louder and louder and louder, and then all of a sudden it stopped. Her NDE took her to what seemed to be a different realm and she awoke to a state of profound confusion over which world was real and which was a dream:

'When I heard my heartbeat stop, I seemed to be like out in space, and I appeared to myself as a light, and there were many, many, many other lights out there with me. It was very, very peaceful, extremely peaceful. It was the first thing I realized about it. It was so peaceful and so restful and quiet. I knew I had to go back. I don't specifically recall someone telling me I had to go back, and I can remember being disappointed.

'Then I remember being above my body in the delivery room and seeing doctors and nurses. And I just went into my body at that point; I don't know how. I don't remember that there was any specific way I went in. I was back into my anesthetized state, I suppose. I didn't wake up for a while.

'When I did wake up, I was terrified, because I couldn't separate reality from what I call my "dream." I knew it wasn't a dream, but I didn't know what else to call it, so I referred to it as a dream in my mind. But I had a terrible time, just an awful time with this. I told my husband the following day, even though I was nursing my baby, and the nurses and doctor would come and go, my husband came, my parents came, his parents came. But I still felt that this was a dream, that reality was the dream and that what I had experienced was reality. This is where I had my problem.

'For weeks, it was just awful, and I finally told my husband on the second day about it, and he said it was just the anesthetic. I said it was different, it was profound. You just can't conjure up words to describe it. I kept telling myself that this was reality, and the other was the dream. And I thought I was losing my mind for quite a while, and this was why I had to suppress this, because every time I would begin to think about it, I had to stop it and push it out of my head and keep saying, "This is real, this is real, not what you experienced." And so I never spoke of it again. I had months of this, where it would come into my mind and I would push it away because it was so frightening to me – not what I experienced, but the aftermath of it just terrified me.

'I was really psychologically upset about this, and it took me a long time to get over it. I didn't know that I had really come near death, but I had trouble separating the two realities. What was happening to me in the hospital, with the baby and the people coming, was the dream; but the other was reality, and that was where I was having the problem. The experience itself was very pleasant, but the sorting out between reality and my experience was the problem and I felt like I had to try to bury the experience to keep my sanity.

'Never having heard of near-death experiences, I could not understand what I had gone through. If it happened today, I would enjoy it. Now I realize that what I experienced was like heaven, like what happens to you after death, but I don't really know that as an absolute yet. It does give you a little security there. I believed in a hereafter anyway, but it's kind of like a backing up of it, I suppose, a little proof.'

Still others have no anger, depression, or confusion themselves, but feel ridiculed or misunderstood when they talk about their near-death experiences. Edith had an NDE at age thirty-eight when her stomach ulcers ruptured. She described her experience as blissful in contrast to the excruciating pain of being in her body:

'The next thing I became aware of was that I was in a new environment. I was suddenly *not* in the bed in the hospital with all the medical paraphernalia restricting my anatomy. But I could see the entire room, complete with a delicate cobweb up in a corner on the ceiling and some cracked plaster over the window. I felt not a twinge of apprehension or fear attached to this unusual situation. I was thoroughly enjoying this whole experience. As I looked around and scrutinized my body from on high, I became aware of an incredible light. This was *not* your everyday beam of sunlight, or a hundred-watt light bulb shining forth, or a roaring fire, or a host of candles. It was not a celestial explosion in the midnight sky. It was radiant, "in sparkling raiment bright." It was warm. It possessed unearthly peace and radiant splendor. Nothing on earth has its color. There are not words to describe the depth of its visionary beauty. This is a place of total love and a place where ultimate security exists, forever.

'From somewhere within the light I felt the presence of another. I looked all around me, but was unable to identify exactly what it was. It was certainly not human, as I know the meaning of the word. It had no definable form; it was not animal, vegetable, or mineral. Instinctively, however, I knew I had nothing to fear from this presence. I felt even more peaceful and secure. As if from all around me, a voice said, "You are safe here."

'I was overcome, because I could not see where it was coming from. Then the voice said that I could stay there forever or I could return to my body and go on with my life; it was my decision. I did not want to go back to all that physical pain. I did not want to go back to unhappiness, conflict, and stress. I liked it there, and I did not want to give up that peace, serenity, and security. I looked down at the body in the bed, with all its artificial appendages; all the more I wanted to stay where I was. As I watched, I began to see my life before me, instantly, on a huge television screen and in living three-dimensional color. It was all there – not a detail was missing – even things I had not thought of for years. The tug-of-war within me was the most intense thing I have ever felt: *go – stay . . . stay – go!*

The voice knew my decision without my having to verbalize it. Because of my husband and my two children, I knew I had to return to my body, no matter what that entailed. With all the heavenly gentleness one can imagine, the voice softly added, "You will go back and you will hold your family together; you will be its cement."

'Then, there I was, in a big, blissful sling, being lowered ever so gently, ever so slowly back into my body. I felt a rousing *thud*! As if by reflex, my eyes opened and I quickly scanned the room. There was *no* question; I was back in my body.'

But after she returned she was ridiculed, misunderstood, and threatened with psychiatric intervention if she mentioned her NDE:

'I was exhilarated! I wanted to tell my story before my insides burst. What a magnificent wonder I had seen! I had to try to describe the light with its infinite power and grace. I thought to myself, "How lucky I am to have brought a piece of it back with me; I will always have it inside."

'A nurse came in to check my blood pressure. I looked at her and began to tell her of my experience. She listened until I was finished. As she unwrapped the cuff she said, "Well, dear, that is very interesting, but you are very ill and you had an hallucination." I thought she really had not understood what I had tried to communicate.

'A little while later a second nurse came in and I told her my story. When I finished, she told me that the drugs they were giving me often caused people to have strange dreams. But I knew what I had seen; I knew what had happened to me, and it was not a strange dream! How could anything so vivid and so real be a dream? I thought I had better wait a longer time before I told anyone of my experience again.

'When the night shift came on, I tried for the third time. When I finished this time, the nurse told me in a cold, matter-of-fact manner that if I continued to talk that way a psychiatrist would be called in. With that pronouncement, I became very frightened. I figured that if the medical profession thought I was crazy, I had better keep my silence about the whole affair. I realized that the best thing I could do was to hold on to the light, never let it go, but keep silent, very, very silent. And so I did.'

Other people find it hard after an NDE to understand and maintain the interpersonal boundaries that are expected in our culture. Having experienced a sense that we are all interconnected, they sometimes reach out in ways that other people may feel is inappropriate. Joe Geraci, the police officer who became a teacher after his NDE, was repeatedly chastised by his principal for 'unprofessional' behavior when he reached out to his students to help them with their personal problems. Alex, a twenty-five-year-old man who had struggled to find his way in life after his NDE, appeared at my doorstep one day with his suitcases and his golden retriever in his car. He had heard me speak about NDEs and had driven two hours to my hometown, expecting me to welcome him into my household for as long as it took to work through his post-NDE issues. Fortunately for my marriage, my training as a psychiatrist included how to set healthy boundaries. Over a lingering dinner at a quiet restaurant, I listened to his problems and concerns and was able to set him up with a counselor in his area who was familiar with the complexities of NDEs, and with a local support group of experiencers so he could talk openly with others.

And this, I realized, was what was needed: more caregivers who would take NDEs seriously as real events, and support groups to counter the devastating feelings some experiencers have of being alone and confused.

During the 1980s public awareness of near-death experiences spread. And as my medical colleagues learned of my interest in NDEs, some of them began referring patients to me for psychiatric treatment for the difficulties they were having as a result of their experiences. After limited success working with each of them in individual psychotherapy, I decided to put them all into a group together. I soon found that they got more support and understanding from one another than from me. The group quickly evolved from group psychotherapy into an open-ended support group, open to anyone, which has continued to meet monthly for more than forty years. Many of the participants brought their family members to the support group. Kenny was a teenager whose heart stopped when he was electrocuted by a freak spark jumping from a high-voltage power line. He had an NDE with both heavenly and hellish visions, and felt he had been saved by Christ and sent back with a mission. His parents had brought him to see me because he felt estranged from his school friends, who didn't understand why he had changed. I brought Kenny to the support group, and he in turn brought his parents so they could understand that his problems were not unique. Long after Kenny himself stopped coming to the group, his parents continued to attend.

In the three decades since Kenny's participation in that group, he continued to wrestle with the aftereffects of his NDE. He now sums up that struggle in these words:

'I've been through a lot of ups and downs since then – some good, some bad. Over the years, I've really discovered the empathic side of me. I know my true gifts lie in the emotional side of humanity and I have a strong ability to comfort and educate when people are at their worst. I do believe the electrocution has shaped my life. I know my life has purpose and I was spared to do something bigger than me, whether it's helping people as a practitioner or just being available to others.'

Because many experiencers have problems reentering their daily lives after their NDEs, in 1984 the International Association for Near-Death Studies (IANDS) sponsored a five-day workshop on helping experiencers. I co-chaired the workshop with Barbara Harris Whitfield, then an IANDS board member who had gone through an NDE herself. We invited thirty-two participants, half caregivers (doctors, nurses, psychologists, social workers, clergy) and half experiencers – and some participants who fit into both categories. The group explored a wide range of therapeutic strategies and techniques that would avoid treating experiencers either as diseased patients or as helpless victims. We came up with general guidelines that focused on three goals. First was helping the experiencers come to whatever understanding of the NDE seemed most helpful to them. Second was respecting the power of the experience as a catalyst for change. And third was focusing on the experiencers' goals going forth in their lives. The workshop also came up with a range of specific techniques for working toward these goals. A key component was the peer support group.

It took another decade for the healthcare establishment to recognize the importance of experiences like NDEs in people's lives. In 1994, the American Psychiatric Association's Diagnostic and Statistical Manual (DSM-IV) acknowledged for the first time that such experiences can cause significant turmoil and lead people to seek help. The DSM-IV included a new category called 'Religious or Spiritual Problems,' embedded in a class of problems that may become the focus of professional attention, but that are not in themselves mental disorders. 'Religious or Spiritual Problems' included, for example, 'experiences that involve loss or questioning of faith . . . or questioning of spiritual values.' In a companion article explaining this new category, the authors gave as one example near-death experiences, using examples from my patients to illustrate that difficulties like anger, depression, and a sense of isolation can frequently arise after NDEs.

Over the years, I've developed some additional guidelines for hospital staff – primarily doctors, nurses, and chaplains – who are likely to encounter patients who have had near-death experiences. Family and friends of experiencers may find them useful as well. They include, first, listening to the experiencer's account of the NDE, without pushing for details or trying to explain or interpret what happened. A second guideline is reassuring the experiencer that NDEs are normal and common, while acknowledging their profound personal impact. A third is encouraging the experiencer to explore any changes in attitudes, beliefs, or values, and how those changes may affect the experiencer's life. It can also be helpful to ask specifically about any anger, sadness, or confusion, and whether there is anything about the NDE or its aftereffects that they find confusing or puzzling or hard to understand or that upsets them.

My colleague Marieta Pehlivanova and I recently studied the types of help and support experiencers seek out in order to help them process their NDEs, the barriers to seeking and receiving help, and whether and how these efforts are beneficial. We found that two-thirds of the experiencers who sought help waited more than a year after the NDE before they did so. When they did seek help, it was for a variety of reasons. The most common reason was because they were struggling with the aftereffects of the NDE, and the second was some troubling feature of the NDE itself. The third most common reason was because they were experiencing problems with other people as a result of the NDE.

One-third of the experiencers who sought help saw mental health professionals. Smaller numbers sought help from spiritual counselors, medical professionals, organizations such as IANDS, online resources such as message boards and social media groups, and religious professionals. One-quarter received individual psychotherapy or counseling. Smaller numbers received hypnosis, meditated, sought medications, went to psychic healers, or attended family counseling, group counseling, or self-help groups. Others engaged in yoga or other body movement therapies.

The good news from this study is that three-fourths of experiencers who sought help were able to find it and felt that it had been a positive experience. The most common benefits they mentioned receiving were insight or new perspectives on their difficulties and validation of their NDEs and their reactions. Others said they benefited from having a safe place to share their thoughts and feelings, being helped to make sense of their NDEs, and receiving emotional support. The bad news is that one-fourth of experiencers who felt the need for help never got it. The most common reasons they mentioned were not knowing that help was available, fearing they would be thought crazy, and fearing they would not be believed.

Today there are many support groups that hold regular meetings for experiencers and the general public. There are more than fifty IANDS-affiliated groups in various cities across the United States and twenty more in various countries around the globe. They provide understanding and information to those who have had NDEs, and offer education and discussion of near-death experiences to the general public in a supportive, accessible setting. In addition, for those who do not have a support group nearby, IANDS provides internet-based support groups through the IANDS Sharing Groups Online, small group discussions emphasizing the sharing of personal experiences in a safe, confidential, and caring environment.

In working with experiencers and their families, I found that troublesome aftereffects are not limited to the survivors of NDEs themselves. Elaine Sawyer complained about Tom's neglect of their family's material needs and of her own personal feelings after his NDE. Family and friends often have difficulty understanding and adapting to experiencers' changes in values, attitudes, beliefs, and behavior after NDEs. The long-term success of family relationships after an NDE can be determined both by how the experiencers integrate the NDE into their lives and by how family members view and accommodate the experiencers' new identities. Researchers in the United States and in Australia have found that marriages in which one partner has had an NDE are less well-adjusted and less stable than they were before the NDE, with 65 percent ending in divorce. This marital instability is usually attributed to problems communicating about a couple's challenges, disagreements on roles, and dissimilar values and goals.

It can be particularly difficult for parents to understand and accept the aftereffects of NDEs in their children. Kenny's parents were confused when their popular, outgoing son lost interest in high school sports, rock music, and socializing with his friends and became focused instead on developing life goals that felt more meaningful to him. Their struggle to understand what he and they were going through kept them coming to the experiencer support group long after Kenny himself stopped.

At one point, I was asked by a pediatric surgeon at my hospital to speak with a mother whose six-year-old son was having open-heart surgery. The operation was required to repair a hole in his heart that he'd had since birth, which was now causing him increasingly irregular heartbeats and occasional difficulty breathing. The night before surgery, his heart had stopped, but then began beating again on its own as the doctors were preparing to shock him. The next morning, when he was being wheeled into the operating room, his mother seemed agitated, and the cardiac surgeon asked for a psychiatrist to speak with her urgently.

When I entered Bobby's room, his mother, Ginger, was sitting in a chair next to his empty bed, fidgeting with a tissue in her hands. She looked up when I knocked on the open door and walked in. I introduced myself and explained that her son's surgeon had asked me to talk with her while Bobby was in surgery. She nodded and then looked down at her hands. She was not crying, but sniffling softly.

'I imagine this all must be pretty scary for you,' I suggested. 'I'd be pretty worried if my son were having heart surgery.'

'It's all so confusing,' Ginger said, stuttering slightly. 'The years of wondering whether he'd be okay, wondering why this happened, whether I'd done something wrong.' She paused, rolling the tissue around between her hands. 'And then the decision to go ahead and have the surgery . . . And then, last night . . .' She shook her head, as if trying to erase the memory.

'Last night?'

'You know that his heart stopped?' she asked, looking up at me for the first time. I nodded, and she quickly looked back down at her hands before I could say anything.

'What happened?' I asked.

'He came out of it, but it made me more worried about the operation this morning . . . And then this morning . . .'

'This morning?'

She kept looking at her hands. 'Just before they came to get Bobby for surgery, I said to him . . .' Her voice cracked a bit, and she swallowed hard. 'I said, "Let's put our hands together and pray that everything will be all right." And Bobby looked me in the eye and said with a big smile, "No, Momma, we don't have to do that."'

She swallowed hard again, her hands still fumbling with the tissue, and then went on. 'He said that we don't have to put our hands together to pray. I . . . I was confused and annoyed, because I was worried about his operation, and here he was giving me lip. So I said to him, "Who told you that?" And he kept staring straight at me and said, "Jesus told me, last night." He scared me, Doctor.'

Ginger choked up and stopped talking. I placed a hand gently on her forearm and she looked up at me, her eyes searching for something.

'I can see how upsetting that was,' I said, nodding. 'What else did he say?'

'He said that Jesus told him the operation was going to be fine, and that his heart would be fixed. And then he said that he put his hands together and asked Jesus if they should pray, and he said Jesus smiled and told him he didn't have to put his hands together. Jesus told him that all he had to do was say the prayer in his heart and God would hear it.'

She paused, still searching my face, and then continued, 'I didn't know what to say, so I just squeezed his hand and didn't say anything. Bobby doesn't talk like that. It was like someone else was talking out of his mouth . . . I was frightened.'

I nodded again, and tried to sound reassuring. 'That must have been scary to hear Bobby talk like that. But it's not unusual. People whose hearts stop, or who are facing some crisis like open-heart surgery, often say they see Jesus or God.'

I felt her forearm relax a bit under my hand, and I went on. 'I know it's frightening, because we don't really understand it. But people who have these experiences are usually fine. They're not upset by it. They usually feel calmer, and this may help Bobby feel more relaxed about his operation. It doesn't mean he's crazy or has anything wrong with him. It just means he was scared of his heart problem and the surgery, just like you are. His experience last night helped him face it.'

Ginger took a deep sigh and nodded. 'But . . . will I get my baby back?'

I smiled and nodded. 'He'll be back, with a stronger heart. And maybe more faith that he's in good hands and that things will work out.'

She smiled and took a deep breath. 'Thank you, Doctor,' she said. 'I'll be okay.'

I paused, trying to assess whether she believed that, and then asked, 'Do you want me to come back later?'

'No, no,' she said quickly. 'As long as the surgery goes

well, I'll be fine.' She paused, then added, 'I just need to sort all this out.'

'Would you like to speak perhaps with the hospital chaplain about this?'

'Maybe,' she said, hesitantly. 'But maybe I'll just speak with my pastor when we get home.'

I gave her my card and told her not to hesitate to call me if she wanted to talk further, either while Bobby was in the hospital or after they'd gone home. She thanked me and I left, wondering whether or not I should mention her to the chaplain. I decided not to, but to leave it to Ginger to ask for help if she wanted it. I wrote a brief note in Bobby's chart, saying that I'd met with his mother to discuss her concerns about the operation, and that I'd be happy to see her again if she or the staff thought that would be helpful.

It is not unusual for family and friends to find that *their* attitudes, beliefs, and behaviors change as a result of intimate exposure to experiencers. And the same is true, I found, for near-death *researchers*. Many times when I left home to visit experiencers or attend an NDE conference, my wife, Jenny, would wonder whether she would get back the same husband she had sent off. And I must admit that sometimes I wondered that as well. I knew that immersing myself in near-death experiences was pushing me to grow and changing my view of the mind and the brain and of who we really are as human beings.

....

A New View of Reality

My high-school girlfriend Jenny was awakened around four a.m. by her mother's alarmed voice: 'Something's wrong with your dad! I can't wake him up. I need your help!' She followed her mother to her parents' bedroom, where her father lay motionless on his back, except for an occasional low gasp. Then the gasping stopped. Jenny's Red Cross Junior Lifeguard training kicked in, and she started doing mouth-to-mouth resuscitation on her father while her mother phoned the doctor. She kept it up for a half hour until the doctor arrived and pronounced her father dead.

Twenty-five years later, my mother-in-law, Alice, was visiting Jenny and me and our kids for the New Year's holiday. We had awakened our preteen son and daughter to watch the Times Square ball drop on TV. It was a cozy time as we snacked and watched the freezing crowds in New York celebrate the new year. After the kids had gone back to sleep, Alice, Jenny, and I sipped champagne and reflected on the past year and our hopes for the new one. After a silence, in which she may have been remembering past New Year's Eves when she and her husband would wake their kids to watch the ball drop, Alice asked, 'Did I ever tell you about the dream I had the night before Jimmy died?'

Jenny and I looked at each other, and then Jenny said, 'No-o-o.' Alice then related a dream of finding herself in a dark room and being aware that there was a man there with her. A door opened, through which brilliant white light poured. The man started to walk through the door, and Alice wanted to follow but couldn't. She was not afraid, and knew that the man was going to be okay, but wanted to go with him. Then he walked out the door and disappeared into the light, leaving her in darkness. She woke from the dream and said to herself, 'That must be what it's like to die. You walk through a door into the light, leaving others behind. I'll have to tell Jimmy about this in the morning.' And then she fell back asleep. 'But I never got the chance,' she now told us. 'He died before I could tell him.'

Both Jenny and I were stunned that she had never told us this dream before. It seemed like such an important part of the story of her husband's death, and yet she now mentioned it almost as an afterthought. And as I listened, I could only think how comforting it must have been for Alice . . . and how reluctant she'd been to share the experience for a quarter century. It made me realize that spiritual experiences around death are often too personal to share, and therefore may be much more common than we know.

It also made me acknowledge how readily I took in Alice's account without blinking an eye. I thought back on how shaken I had been when Holly mentioned the spaghetti stain on my tie, and now, decades later, my mother-in-law relating what seemed to be a premonition of her husband's death struck me as perfectly plausible.

The intervening years in medical school and psychiatric training and the decades studying near-death experiences exposed me to a variety of situations where brain function didn't seem to explain our thoughts and feelings. I had come to appreciate that some of the things I'd been taught about the mind and the brain were assumptions rather than facts. The idea that our minds – our thoughts, feelings, hopes, fears – are produced solely by our physical brains is not a scientific fact. It is a philosophical theory proposed to explain scientific facts. And it is only one of many such theories – some of which may offer better explanations of thoughts and feelings. Over the decades I had become comfortable entertaining a variety of theories about the mind and the brain, using different models for different tasks.

Experiences like NDEs seem to me to involve both the physical brain and the nonphysical mind. We can choose to focus on the physical brain and explore chemical and electrical changes associated with NDEs. Or we can focus on the nonphysical mind and explore feelings of peace and love, out-of-body perceptions, and encounters with deceased loved ones. Both aspects – the physical and the nonphysical – are there, and we can see either one by changing our focus. We can look at NDEs as a function of the physical brain or as a function of the nonphysical mind – but neither of those perspectives by itself provides a complete description of the experience.

Almost twenty years ago, neuroscientist Andy Newberg at the University of Pennsylvania measured blood flow in the brains of Franciscan nuns as they prayed. He found increased activity in certain parts of the brain. When he showed these brain scans to his fellow neuroscientists, their response was purely physical: 'So those are the parts of the brain that make them think they're talking with God!' But when he showed the same brain scans to the nuns, their response combined the physical and nonphysical: 'So those are the parts of the brain God uses to talk to me!' As Andy summarized it: 'Skeptics used my findings to conclude that religious experience was nothing more than a neural con-fabulation within the brain, and religious practitioners cited my work to confirm that human beings are biologi-cally "hardwired for God."'

So when experiences like NDEs can be interpreted in more than one way, how do we decide which model to use? Are NDEs the result of changes in the physical brain, or are they experiences of the nonphysical mind? Do we have to choose, or can we see both?

It seems plausible to me that NDEs may be triggered by electrical or chemical changes in the brain that permit the mind to experience separating from the body at the moment of death. There is no inherent conflict between a physical and a nonphysical understanding of NDEs. The physical and the nonphysical are different levels of explan-ation or description. It's like saying that my desk is mahogany – a physical description – and that my desk is a legacy from my grandfather – a nonphysical one. They are both correct, but neither by itself gives a complete description of my desk. So, too, physical and nonphysical descriptions of NDEs may both be right, but neither by itself gives a complete picture of NDEs.

In our daily lives, our physical and nonphysical parts seem to work together as one. Changes in the physical body may lead to profound changes in the nonphysical mind. Several years ago, when some of my friends were retiring, I was envious of them but couldn't imagine retiring myself. I was enjoying my clinical practice treating patients with psychiatric problems, I was enjoying teaching medical students and psychiatric trainees, and I was enjoying studying near-death experiences. While those activities never totally defined who I was, they were such a large and fulfilling part of my life that I couldn't imagine what would fill the hole if I stopped doing them.

And then my hip deteriorated and I needed surgery to replace it. After the operation, I was largely confined to bed for several weeks, followed by more weeks of physical therapy before I could return to work. To my great surprise, those weeks of forced absence from work opened up unexpected new opportunities that I found every bit as fulfilling and exciting as my job as a psychiatry professor. It gave me more time to spend with my wife, and allowed me to see different ways I could share what I'd learned about NDEs with the wider world. In a very real sense, the physical event of getting a hip replacement led to profound nonphysical changes in my attitude, including a readiness to retire.

And just as physical changes in our bodies can lead to changes in our minds, so too changes in our nonphysical thoughts and feelings can lead to changes in our physical bodies. Everything we think and feel produces changes in our brain. When you feel carried away by a dazzling sunset or savor a chocolate truffle, or when you feel joy at helping someone else in need, those feelings are all associated with electrical and chemical changes in your brain. MRI scans of the brain show that meditation, a mental practice of focusing the nonphysical mind, changes your physical brain over time by decreasing the size of brain areas that react to stress. And both functional MRI brain scans and EEGs of experiencers show that meditating on their NDEs changes their physical brains by increasing electrical activity and blood flow in areas involved in positive emotions and mental imagery. MRI, PET, and SPECT scans of the brain in patients undergoing psychotherapy show that changing thoughts through psychotherapy, another nonphysical process, changes the physical brain by decreasing blood flow and metabolic activity in regions associated with anxiety or depression.

Although our physical brain and nonphysical mind seem to work as one unit in everyday life, people who have had NDEs consistently say that their experience of being awake and aware while their brains are impaired convinces them that their minds can act independently of their brains at times, and are not just the product of their physical brains. And that leads them to believe that their mind or consciousness may *continue* after the physical body dies. What are those of us who have not had NDEs to make of all this? Since most experiencers say that their NDE can't be adequately expressed in words and that they have to resort to metaphors, can we really be certain about how to interpret what they experienced?

I expect that many of you are like me. We live our lives based on the evidence of our eyes and ears and our logical deductions from that evidence. We haven't received divine revelations that tell us what's true and what's not. Most of the people I studied who had NDEs feel certain they know the truth because of the very personal evidence of their experience. So where does that leave the rest of us who haven't had NDEs, and don't have that certainty? How can we evaluate the claims of experiencers about the truths that were revealed to them in their NDEs?

I've come to accept that we don't have all the answers. Uncertainty and ambiguity no longer scare me because studying NDEs has made me more comfortable with not having all the answers. A couple of years ago, I was lying down one afternoon, not asleep but very relaxed. I was in a rather dreamy state when I had the feeling that my body was getting larger and larger. At first, there was no particular emotion associated with that sensation, but as I kept growing, it seemed that I had become far bigger than the planet Earth. As I continued to expand through the universe, reaching toward the distant stars, I suddenly realized that the atoms that made up my body had not grown in size, but I kept getting larger because the distance between my individual atoms was increasing. With a shock, I recognized this experience as the one I'd had in the terrifying dream decades earlier, the night before I was to give a talk to the American Psychiatric Association. And just as with the dream from years ago, I knew that it was just my imagination and my emotions. But whereas that dream years ago had been terrifying, *this* experience was blissful. Instead of feeling panic as I flitted back and forth between my rapidly separating atoms, I was enjoying the freedom of expanding into the universe. I did not feel the need to keep the atoms of my body together, but relished the feeling of exploring the vastness of the cosmos.

I emerged from that experience feeling rejuvenated and more alive – not to my body shaking and drenched in sweat, as I had decades earlier. Of course, I am much older now, and this time I was not under the pressure of giving a public lecture. But even so, this experience felt *so* different from the earlier terrifying dream that had the same content, the same plot. I think the two experiences were different because I was different. The cumulative effect of listening to experiencers over the intervening years has helped me come to terms with the unknown and feel comfortable with the unexplained.

For forty years, I have seen how near-death experiences profoundly affect the experiencers themselves and people who come in contact with them. And I've seen the effects they've had on me as a researcher. But what about people who haven't had such long-term involvement with near-death experiences? Can they be affected by NDEs as well?

In fact, it turns out that some of the changes in attitudes, values, and behavior that we see in experiencers have shown up in other people who just learn about NDEs secondhand. Psychologist Ken Ring calls these aftereffects a 'benign virus' that other people can catch from experiencers – or from other people who have been similarly infected. There is a growing number of clinical reports in the medical literature showing the power of learning about NDEs to provide comfort, hope, and inspiration to persons who have not themselves had these experiences.

Five studies among college students confirm these secondhand effects. One at Miami University of Ohio found that more than 80 percent of undergraduate students in a sociology class that studied NDEs felt more compassionate concern for others and greater feelings of self-worth both at the end of the semester and in a follow-up a year later. Another study at Montana State University found that nursing students who completed a course on NDEs had less fear of death, a more spiritual orientation, and a greater sense of purpose in life. Two separate studies at the University of Connecticut found that undergraduate students in a psychology class on NDEs had greater appreciation of life, more self-acceptance, and more compassionate concern for other people. They also reported a greater sense of spirituality, less interest in material possessions, and less fear of death.

And a study at Massey University in New Zealand compared students assigned randomly to one group that viewed online educational material about NDEs to a second group that did not view the NDE material. Those who were exposed to the NDE information had greater appreciation for life, greater spirituality, and a more positive attitude toward death; and they had less anxiety about material possessions and achievement.

Information about NDEs was also included in the health curriculum of a high school in Kentucky – including one teacher talking about her own NDE when she had an aneurysm. The teacher described leaving her body and feeling peaceful, and after the experience feeling a renewed vigor for living and no longer fearing death. A preliminary report of the effects of teaching about NDEs indicates positive emotional and behavioral changes in the students. Thus all six studies carried out so far concluded that educating high school, college, or nursing students about NDEs had positive effects on them.

After decades of teaching doctors, nurses, hospital chaplains, and other healthcare workers about near-death experiences, I am thrilled to see that awareness of NDEs is beginning to influence the practice of healthcare as well. Many medical and nursing schools now include information about NDEs in their curricula. In recent years, this information has inspired new approaches to patients, as healthcare providers have become more sensitive to the frequency and effects of NDEs in their patients.

Studies have shown that introducing information about NDEs into the treatment of suicidal clients who did not respond to traditional therapy can lead to a dramatic decrease or elimination of suicidal thoughts. Other research has shown that information about NDEs can reduce the suffering of grieving individuals, leading to less anxiety, anger, and blame, and help them feel engaged in life again. So near-death experiences appear to be having a rippling effect throughout society in helping people address their concerns about death and their ability to enjoy life and feel compassion for one another.

There is growing evidence that spreading awareness of near-death experiences and their implications not only *can* influence other people for the better, but *is* doing so. Joe Geraci, who bled out after surgery at age thirty-six, summarized this point for me:

'I think our society can be very negative: "Don't do this. Don't do that," a very black-and-white, closed system. But if people just loved and cared, there'd be no need to concern ourselves with all the "don'ts." There just wouldn't be. I know that sounds very idealistic and unpractical. But I believe love can be just as infectious as hate.

'And to do that, people have to start somewhere. On a small scale, just me telling you about my experience, and someone reading what you're going to write. It multiplies quickly. And I'm not the only one that has had this experience. There are thousands of us all around the world. Multiply my story by a thousand and you'll see how quickly it can grow! It can be done. In fact, it has already started.'

Life before Death

Much of the public interest in near-death experiences is related to the hope that they may tell us about life after death. And indeed, most experiencers are convinced that some aspect of us does continue after death. But they also consider of equal importance the lessons they bring back from their NDEs for life *before* death. The experience often gives them a new outlook on what makes this life purposeful and meaningful. I used the title *After* for this book to reflect this mixed focus. It refers to what happens to people *after death* – but also to what happens to people in this life *after an NDE*. As I understand near-death experiences, they are ultimately not about death, but about transformation, about renewal, and about infusing our lives with purpose right now.

My hope is that this book will help move the conversation about NDEs beyond what they suggest about mind and brain and a possible afterlife. I'd like to see the conversation expand to include the more critical issues of life here and now. We may well find ourselves somewhere else after we die, but we are *here now*. After working for a half century with people who had near-death experiences, I've learned a number of lessons about NDEs and what they imply for us right now.

The first lesson I learned is that *NDEs are common experiences that can happen to anyone*. Most researchers estimate that between 10 and 20 percent of people who come close to death report NDEs – or about 5 percent of the general population. Numerous studies over the past forty years have failed to find any variable that can predict who will have an NDE. They happen to men and women, to people of all ages, all religions, and all ethnic groups. They are not by any means rare events, nor are they experiences that only certain kinds of people have. What does this mean for those of us who haven't had NDEs? The fact that NDEs are common experiences that can happen to anyone means that sooner or later you're going to meet someone who's had one – if you haven't already. If one in twenty Americans has had an NDE, it is likely that at least one person in your extended family or in your workplace or in your classroom has had one.

The second lesson is that *NDEs are normal experiences that happen to people in exceptional circumstances*. Memories of

NDEs look like memories of real events, not like memories of fantasies or imagined events. Our brains process NDEs like events that really happened, not like dreams or hallucinations. Numerous studies have failed to find any link between NDEs and any type of mental illness. In fact, several studies suggest that NDEs may offer some *protection* against the development of mental illness after a close brush with death. What does that mean for the rest of us? The fact that NDEs are normal experiences and not signs of mental illness means that experiencers do not need to be referred to a counselor or mental health professional just because they had an NDE. What they need from the rest of us is reassurance that they are 'normal,' validation that their experiences were real, and the opportunity to share their experiences and to reach their own understanding of them.

The third lesson is that *NDEs usually lead to a number of profound and long-lasting aftereffects*. Whether those aftereffects are positive, like enhanced enjoyment of life, or negative, like difficulty fitting back in to one's job or lifestyle, it is usually helpful to acknowledge and address the changes and what they mean for the experiencer's life and relationships. Although most experiencers cope with these changes by themselves, family and friends as well as healthcare workers should be aware of these aftereffects and alert to indications that the experiencer may need help. If you are close to an experiencer, you may need to be aware of changes in your relationship and perhaps help the experiencer decide what life changes he or she wants to make in order to fit the aftereffects into their daily life.

The fourth lesson is that *NDEs reduce fear of death*. Most people expect dying to be a terrifying experience. However, experiencers almost universally report that having an NDE greatly diminished their fear of death and dying – and many say their fear was completely erased. This is true for experiencers whether their NDE was a typically blissful one or a rare frightening one. Knowing that NDEs reduce experiencers' fear of death may make you think differently about your own death. Knowing that the process of dying is usually peaceful – if not blissful – may mean you don't need to be afraid of dying. It may also make you worry less about loved ones suffering as they die. You should not expect, however, that it will stop you from grieving. The death of a loved one is still the loss of a relationship and a shared history. Even if you are not concerned about their suffering, you may still feel the pain of their loss. After their NDEs, experiencers still grieve when others die.

One consequence of their reducing fear of dying is that NDEs paradoxically also reduce fear of *living*. Many experiencers say that because they're no longer afraid of death, they no longer feel there's as much to lose as they'd thought before the NDE. They no longer feel the need to maintain such tight control over their lives, but can feel free to take some risks. The fact that being less afraid of death leads experiencers to feel less afraid of living may encourage the rest of us to open up and enjoy all life has to offer, with less fear of making mistakes.

And that brings us to the related fifth lesson, that *NDEs lead experiencers to live more fully in the present moment*, rather than dwell in the past or dream of the future. I suspect this tendency to live in the present comes at least in part from the experience of almost dying – that is, of encountering a moment they think might be their last. So it makes sense to me that people who have had NDEs might continue to live with the memory of that experience, and therefore try to make the most of each day. People who survived NDEs believe they *were* experiencing their last moments, with no chance to say goodbye or to resolve unfinished business.

If we all thought that this moment *right now* might be our last, how would we act toward our spouses, our children, our friends, strangers we meet on the street – and ourselves? The example of how experiencers, having almost died, try to live more in the present may encourage the rest of us to enjoy life more fully and to live in the moment. And making the most of life, living life more fully in the present, allows us to appreciate not only the joy but also the pain of each experience.

John Wren-Lewis was poisoned by a would-be thief on a bus traveling across Thailand with his wife, Ann Faraday. After a while, Ann noticed with alarm that John was blue around the lips and she couldn't feel his pulse. She managed to get them a ride to a nearby hospital, where the doctors were not hopeful of saving him. Assuming that he had been drugged – a common tactic of thieves in that region – they gave him an antidote to narcotics, oxygen, and an intravenous drip. He revived about seven hours later, after a profoundly moving NDE, in a state of what he called 'eternity consciousness.' He described for me his heightened appreciation for life, which stayed with him for the rest of his life:

'Although I get more pleasure than before from "good" experiences like sunsets, birdsong, great art, pleasant people, or delicious food, I also get as much pleasure from things which in my old state I would have called unpleasant: for example, the Thai hospital room, or a very wet day, or a heavy cold. This last discovery that I could positively enjoy a cold – not merely wallow in the indulgence of a day in bed but really get a kick from the unusual sensations in my nose and throat – was a big surprise.

'Around that time, I found that the tinnitus, the hissing in the ears from which I have suffered for some years, had changed from being a mild annoyance, which I could at best manage to forget at times, to a positively delightful sound, which I welcomed as an old friend whenever it forced itself on my attention. I also actually started to enjoy tiredness and the many minor pains that afflict a sixty-year-old body.'

Understanding that these experiences can lead people to approach every day as if it might be their last may help us to fill each day not only with obligations but also with joy, knowing that there might not be a second chance. As poet Patricia Clafford put it, 'The work will wait while you show the child the rainbow, but the rainbow won't wait while you do the work.' Living fully in the moment doesn't mean never planning for the future or remembering the past. It means being fully in the present *while* you are planning or remembering, and letting yourself get completely engrossed in your experience of the moment.

The sixth lesson is that *NDEs raise questions about the relationship between minds and brains*. In normal everyday life, it often seems as if our brains and our minds are the same thing. But experiencers almost universally report that their thoughts and perceptions are clearer than ever in NDEs while their brains are severely impaired. Furthermore, they sometimes report accurate perceptions of things going on around their bodies, but from a point of view *outside* their physical bodies. Those paradoxes suggest that we need a different model of how minds and brains interact. They suggest that the physical brain may act like a cell phone, receiving thoughts and feelings from the nonphysical mind and converting them into electrical and chemical signals that the body can understand and use. And they suggest that – in extreme circumstances at least – minds *can* function quite well without brains to filter them.

I don't know whether the mind operating separate from the brain is the best explanation for NDEs in which the experiencer sees things accurately from a viewpoint outside their physical bodies. But I don't have any alternative explanation of the evidence. And I don't know whether brains acting as filters of thoughts and feelings, the way eyes act as filters of light waves, is the best explanation for NDEs in which experiencers think and see clearly while their brains are severely impaired. Certainly that model raises further questions about what and where the mind is and exactly how it interacts with the brain. But I don't have any alternative explanation of the evidence. We may eventually come up with another explanation, but until then, minds and brains as separate things, with brains acting to filter our thoughts and feelings, seems to be the most plausible working model.

The questions NDEs raise about how our thoughts and feelings relate to our brains may make you think about whether we are just biological machines, or something more than that. Whether or not the idea of a mind functioning independent of a brain makes sense to you, NDEs should make you skeptical about our current models of how the brain and mind work, and whether there may be more to our thoughts and feelings than electrical and chemical changes in our brain cells.

The seventh lesson is that *NDEs raise questions about the continuation of consciousness after death.* If it is true that minds can function in extreme circumstances without physical brains, then it may be possible that minds can continue to exist after the death of the brain. An answer to what happens after death may be beyond today's scientific methods – or it may be just beyond our scientific imagination. But the scientific answer – if it ever comes – will likely be through indirect evidence, like the trail of bubbles subatomic particles leave in a bubble chamber. I don't know whether some kind of continued consciousness after death is the best explanation for NDEs in which experiencers see deceased loved ones no one knew had died. But I don't have any alternative explanation for the evidence. We may eventually come up with another explanation, but until then, some form of continued consciousness after death seems to be the most plausible working model.

The evidence that under extreme conditions we can perceive beyond what our physical senses see and hear, and that we can remember things our physical brains have not processed, comes not just from NDEs but from a variety of research avenues. So it makes sense to me to live our lives as if this is really the way things are – that we are more than our physical bodies, that some part of us may continue after our bodies stop working, and that we may be intimately connected to something greater than ourselves. And that has tremendous implications for how we live our lives, and for what makes our lives meaningful and worthwhile.

Several years ago, I was invited to the Dalai Lama's compound in Dharamsala, India, to participate in a dialogue between Buddhist scholars and Western scientists on mind and matter. I presented the scientific research on whether consciousness is produced by the brain, focusing heavily on NDEs. Unlike audiences I typically address here in the US, the Buddhist monks were quite familiar with the experiences I described, although they were surprised to learn that scientists were studying them.

More important to me, however, was a comment by the

Dalai Lama himself about the difference between Western science and Buddhism. Both disciplines, he argued, are based on observation and logical deduction, and both give experience precedence over belief in their quest for the truth. But, he added, Western scientists seem to seek understanding about how the world works in order to change and *control* the natural world. That is, the goal of most scientists is to gain mastery over our environment. Buddhists, on the other hand, seek understanding about how the world works in order to *live more harmoniously* with it. In other words, the goal of Buddhism is to coexist with nature rather than gain mastery over it, in order to reduce our suffering. That distinction affected me deeply, and has made me question everything I do as a scientist, why I do it, and what purpose it will serve. It changed my reason for doing research from 'What might we learn from these results about how the world works?' to 'How might these results help reduce suffering in the world?'

These goals of controlling nature and reducing suffering are not necessarily mutually exclusive. Medical scientists regularly study diseases in order both to change the course of an illness *and* to relieve patients' suffering. But the Buddhist perspective suggests that understanding even phenomena we cannot change may help alleviate suffering in the world. Near-death experiences are beyond our control, and there is no reason to think that will ever change. But we can understand them and their aftereffects. And the evidence we have so far suggests that understanding them better – including NDEs in our science and our medicine – can help reduce suffering.

*

What does all this mean for those of us who have not had NDEs? I had a third reason for using the title *After* for this book. The title refers not only to what might happen *after death* and to what happens *after NDEs*, but also to what might happen *after you've read this book*. I would hope that your reflections on my words will not end when you put this book down, but may continue to live on in your thoughts and feelings about life, death, and beyond.

However we understand what causes NDEs, they show us that there is much more to learn than we now know about the mind and its abilities. But more than that, NDEs seem to give people who have them the spark to reevaluate their lives and make changes in how they spend their time and how they relate to other people. They tell us that death is more about peace and light than about fear and suffering. They tell us that life is more about meaning and compassion than about wealth and control. They tell us that appreciating both the physical and the nonphysical aspects of life gives us a much fuller understanding. And the evidence shows that near-death experiences transform the lives not only of people who have them and their loved ones, or the researchers who study them. NDEs can also transform those who read about them and can ultimately, I believe, even help us change the way we see and treat one another. It is my hope that learning about near-death experiences will also give you the spark to reevaluate your life and reconnect with the things that fill your life with ever greater meaning and joy.

Acknowledgments

....

It is abundantly clear to me that I could not have written this book – or enjoyed the career that I have – without the guidance and collaboration of a great many people over the years. And they deserve much of the credit for mentoring and encouraging me along this journey.

First and foremost, I need to express my boundless thanks to the multitude of experiencers who have participated in my research, some of whom have been filling out my questionnaires about their near-death experiences for more than forty years. Many of them have offered insightful comments about this research and suggestions for further investigations. Without the generous sharing of their time, knowledge, and wisdom, none of this work would have been possible. I feel privileged to be able to share what I have learned from them.

I owe tremendous debts to my father, Bill Greyson, who instilled in me from an early age a passion for science and for knowledge based on evidence rather than on beliefs; and to my mother, Debbie Greyson, who taught me from as early an age that nothing we do means much unless it comes from the heart; and to both of them for modeling for me that, whatever you do with your life, the measure of your success is whether you've helped other people.

I am also deeply grateful to the late Ian Stevenson, who showed me how to apply scientific methods to the study of unexplained phenomena, and to Raymond Moody, who introduced me to near-death experiences, as he did for so many other people. I'm also indebted to the pioneers who cofounded with me the International Association for Near-Death Studies (www.iands.org), the organization that put near-death experiences on the map: Ken Ring, who led the way to the rigorous study of near-death experiences, Michael Sabom, and John Audette.

In addition, I want to express my gratitude to those colleagues who have collaborated on and added so much to my near-death research, particularly the late Ian Stevenson, Ken Ring, Emily Williams Kelly, Surbhi Khanna, Jan Miner Holden, Ed Kelly, Nancy Evans Bush, Masayuki Ohkado, Sam Parnia, Peter Fenwick, Barbara Harris Whitfield, Lauren Moore, Marieta Pehlivanova, Rense Lange, Jim Houran, Mitch Liester, Geena Athappilly, Adriana Sleutjes, Alexander Moreira-Almeida, Enrico Facco, Christian Agrillo, Karl Jansen, Evgeny Krupitsky, Jeff Long, Pim van Lommel, Ross Dunseath, the late John Buckman, Debbie James, Cheryl Fracasso, Harris Friedman, the late Chuck Flynn, David Hufford, Jim Tucker, Paul Mounsey, Alan Marty, Nathan Fountain, Lori Derr, Donna Broshek, Jim Council, Karen Packard, Lisa Hacker, Charles Paxton, Claudia Szobot, Charlotte Martial, Héléna Cassol,

Vanessa Charland-Verville, Steven Laureys, and Enzo Tagliazucchi. I was particularly fortunate to have had as collaborators some researchers who had themselves had NDEs, who have kept me from getting sidetracked from the experience into my logical scientific approach – and other collaborators who had not had NDEs, who have kept me from getting sidetracked from my logical scientific approach into the experience.

I am also deeply indebted to my colleagues at the University of Virginia Division of Perceptual Studies (www.uvadops.org) who have critiqued and improved my work, including Ed Kelly, Emily Williams Kelly, Jim Tucker, Lori Derr, Marieta Pehlivanova, Carlos Alvarado, Nancy Zingrone, Kim Penberthy, Ross Dunseath, and Christina Fritz; and to Sue Ruddock, Pat Estes, and Diane Kyser, who helped immensely with the everyday work of my research. The Division of Perceptual Studies, which receives no financial support for research from the university but has pursued rigorous scientific investigations for more than a half century – funded entirely by donations – gave me a safe place in which to explore the unknown world, and continues to do so now for other scholars.

I am also immeasurably grateful to the late Chester F. Carlson and the late Priscilla Woolfan, whose bequests funded my endowed chair at the University of Virginia; to the nonprofit research foundations that have funded my research over the years, particularly the BIAL Foundation (Fundação BIAL), the Institute for Frontier Areas of Psychology and Mental Health (Institut für Grenzgebiete der Psychologie und Psychohygiene), the Japan-US Fund for

Health Sciences, the Azuma Nagamasa Fund, the James S. McDonnell Foundation, the Bernstein Brothers Foundation, and the Fetzer Institute (formerly the John E. Fetzer Foundation); to the support of this research by Richard Adams, Cheryl Birch, and David Leiter; and to the support of my colleagues at the Universities of Virginia, Michigan, and Connecticut.

I also acknowledge my debt to the many colleagues over the decades who panned and picked holes in my research. Their criticisms undoubtedly improved the quality of my work and contributed to a better understanding of near-death experiences.

Few books are written by one person alone, and I had plenty of help with this one. I am deeply indebted to the many experiencers who initially provided the statements that I have quoted in this book. Every word I wrote was read first by my wife, Jenny, who knows me better than I know myself. She made sure my descriptions of my reactions to these accounts were authentic, and that the emotions and tone of voice were genuine. I then ran every word past my talented collaborator Jason Buchholz, who showed me how to bring the story alive and turn the book that I wanted to write into one that others might want to read. Finally, my older and wiser sister Nancy Beckerman painstakingly found and corrected all my blunders and ambiguities, as she has done throughout my life.

I owe undying thanks to the visionary team at Idea Architects, who helped me turn my idea for this book into a reality, particularly Doug Abrams and Lara Love Hardin, whose enthusiasm and passion for my work were critical in guiding me to create this book. Thanks are also due to George Witte, my editor at St Martin's Press, who provided much guidance and practical advice about crafting this book. I am also indebted to Steve Bhaerman, aka Swami Beyondananda, who encouraged me forty years ago to write a book; and to Rebecca Valla, a boundlessly insightful psychiatrist with a heart of gold, who sat me down and helped me understand my introvert's reluctance to expose myself in writing.

Finally, I want to express my eternal gratitude to Jenny Greyson, my compass, my anchor, my life partner, and my best friend for more than a half century, who has been my rock through the ups and downs of near-death research for most of our life together. There is no way I could have developed as a father, friend, psychiatrist, or writer, without her love and support.

Notes

....

INTRODUCTION: A JOURNEY INTO UNCHARTED TERRITORY

13. **Life After Life** . . . Raymond A. Moody, *Life After Life* (Covington, GA: Mockingbird Books, 1975).

14. **NDE accounts from ancient Greek and Roman sources** . . . Jeno Platthy, *Near-Death Experiences in Antiquity* (Santa Claus, IN: Federation of International Poetry Foundations of UNESCO, 1992).

14. **all the major religious traditions** . . . Farnaz Masumian, 'World Religions and Near-Death Experiences,' in *The Handbook of Near-Death Experiences*, ed. by Janice Miner Holden, Bruce Greyson, and Debbie James (Santa Barbara, CA: Praeger/ABC-CLIO, 2009), 159–83.

14. **indigenous populations around the world** . . . Allan Kellehear, 'Census of Non-Western Near-Death Experiences to 2005: Observations and Critical Reflections,' in *The Handbook of Near-Death Experiences*, 135–58.

14. **the medical literature of the nineteenth and early twentieth centuries** . . . Terry Basford, *The Near-Death Experience: An Annotated Bibliography* (New York: Garland, 1990).

14. **themes in these experiences that go beyond cultural interpretations** . . . Geena Athappilly, Bruce Greyson, and Ian Stevenson, 'Do Prevailing Societal Models Influence Reports of Near-Death Experiences? A Comparison of Accounts Reported before and after 1975,' *Journal of Nervous and Mental Disease* 194(3) (2006), 218–22.

1. A Science of the Unexplained

22. Henry lodged the .22-caliber hunting rifle between his legs . . .
Henry's near-death experience and my psychological interpretation of it
are described in John Buckman and Bruce Greyson, 'Attempted Suicide
and Bereavement,' in *Suicide and Bereavement*, ed. by Bruce L. Danto and
Austin H. Kutscher (New York: Foundation of Thanatology, 1977),
90–104.

31. Sigmund Freud compared the mind to an iceberg . . . Sigmund
Freud, 'The Unconscious,' in *Standard Edition of the Complete Psychological
Works of Sigmund Freud, Vol. 14*, ed. by James Strachey (London: Hogarth
Press, 1915), 159–204.

**31. most classroom teachers would not knowingly give better grades
to the most attractive students . . .** David Landy and Harold Sigall,
'Beauty Is Talent: Task Evaluation as a Function of the Performer's
Physical Attractiveness,' *Journal of Personality and Social Psychology* 29(3)
(1974), 299–304.

**36. that experiment was later published in a mainstream medical
journal . . .** Bruce Greyson, 'Telepathy in Mental Illness: Deluge or
Delusion?' *Journal of Nervous and Mental Disease* 165(3) (1977),
184–200.

36. he had written a book while still a medical student . . . Raymond
A. Moody, *Life After Life* (Covington, GA: Mockingbird Books, 1975).

2. Outside of Time

**43. Heim published the first large collection of near-death
experiences . . .** Albert von St Gallen Heim, 'Notizen über den Tod durch
Absturz [Notes on Fatal Falls],' *Jahrbuch des Schweizer Alpen-Club
[Yearbook of the Swiss Alpine Club]* 27 (1892), 327–37.

43. 'What I felt in five to ten seconds' . . . This quote is from the
English translation of Heim's near-death experience, published in Russell
Noyes and Roy Kletti, 'The Experience of Dying from Falls,' *Omega* 3
(1972), 45–52.

**46. Joe Green has raised the question of whether Heim's account of
his fall played a role in Einstein's theory of relativity . . .** Joseph
Timothy Green, 'Did NDEs Play a Seminal Role in the Formulation of
Einstein's Theory of Relativity?' *Journal of Near-Death Studies* 20(1)
(2001), 64–66.

47. 'Time became greatly expanded' . . . This quote is from page 47 of
Noyes and Kletti, 'The Experience of Dying from Falls.'

47. One of those students was a teenage Albert Einstein . . . Ronald W. Clark, *Einstein: The Life and Times* (New York: Avon, 1971), 54.

47. he later described in a letter to Heim's son as 'magical' . . . Albrecht Fölsing, translated by Ewald Osers, *Albert Einstein* (New York: Penguin, 1997), 66.

47. which proposed that *time slows down the faster you travel* . . . His theory of relativity was proposed in Albert Einstein, 'Zur Elektrodynamik bewegter Körper,' *Annalen der Physik* 322(10) (1905), 891–921. (Translated into English by George Barker Jeffery and Wilfred Perrett and published as 'On the Electrodynamics of Moving Bodies' in *The Principle of Relativity*. London: Methuen, 1923.)

47. Joe Geraci, a thirty-six-year-old policeman who almost bled to death after surgery, described this sense in his NDE . . . Joe's NDE was described in Darlene Taylor, 'Profile of an Experiencer: Joe Geraci,' *Vital Signs* 1(3) (1981), 3 and 12.

48. They were less likely in NDEs that might have been anticipated . . . Ian Stevenson and Bruce Greyson, 'Near-Death Experiences: Relevance to the Question of Survival after Death,' *JAMA* 242(3) (1979), 265–67; Bruce Greyson, 'A Typology of Near-Death Experiences,' *American Journal of Psychiatry* 142(8) (1985), 967–69; Bruce Greyson, 'Varieties of Near-Death Experience,' *Psychiatry* 56(4) (1993), 390–99.

49. Jayne Smith had an NDE at age twenty-three during a bad reaction to anesthesia . . . Jayne described her near-death experience in Jayne Smith, '. . . Caught Up into Paradise,' *Vital Signs* 3(1) (1983), 7 and 10; and in Jayne Smith, 'Unconditional Love: The Power and the Glory,' *Vital Signs* 19(1) (2000), 4.

3. THE LIFE REVIEW

52. the 'life review,' in which scenes from the experiencer's past come flooding back . . . Ian Stevenson and Emily Williams Cook, 'Involuntary Memories during Severe Physical Illness or Injury,' *Journal of Nervous and Mental Disease* 183(7) (1995), 452–58; Russell Noyes and Roy Kletti, 'Panoramic Memory: A Response to the Threat of Death,' *Omega* 8(3) (1977) 181–94.

54. he fell off a boat into Portsmouth Harbor . . . This description of his NDE appears on pages 77–78 of Francis Beaufort, *Notice of Rear-Admiral Sir Francis Beaufort, K.C.B.* (London: J. D. Potter, 1858).

55. when a truck he was working under came crashing down on his chest . . . Tom described his NDE in Sidney Saylor Farr, *What Tom*

Sawyer Learned from Dying (Norfolk, VA: Hampton Roads Publishing, 1993).

60. Barbara Harris Whitfield had an NDE at age thirty-two when she suffered respiratory complications . . . Barbara described her NDE in Barbara Harris and Lionel C. Bascom, *Full Circle* (New York: Pocket Books, 1990); and in Barbara Harris Whitfield, *Final Passage* (Deerfield Beach, FL: Health Communications, 1998).

63. a major tool for counselors working with people at the end of their lives . . . David Haber, 'Life Review: Implementation, Theory, Research, and Therapy,' *International Journal of Aging and Human Development* 63(2) (2006), 153–71; Robert N. Butler, 'The Life Review: An Interpretation of Reminiscence in the Aged,' *Psychiatry* 26(1) (1963), 65–76; Myrna I Lewis and Robert N. Butler, 'Life-Review Therapy: Putting Memories to Work in Individual and Group Psychotherapy,' *Geriatrics* 29(11) (1974), 165–73.

4. GETTING THE WHOLE STORY

66. I interviewed almost 1,600 patients . . . Bruce Greyson, 'Incidence and Correlates of Near-Death Experiences in a Cardiac Care Unit,' *General Hospital Psychiatry* 25(4) (2003), 269–76.

69. Bill Urfer, a forty-six-year-old businessman, related to me his difficulty describing the NDE . . . Bill described his NDE in Harry Cannaday (as told by Bill Urfer), *Beyond Tomorrow* (Heber Springs, AR: Bill Urfer, 1980).

70. 'Silence is the language of God' . . . This quote appears on page 134 of Igor Kononenko and Irena Roglič Kononenko, *Teachers of Wisdom* (Pittsburgh: RoseDog Books/Dorrance, 2010).

71. interviewing patients who were hospitalized after attempting suicide . . . Bruce Greyson, 'Near-Death Experiences and Attempted Suicide,' *Suicide and Life-Threatening Behavior* 11(1) (1981), 10–16; Bruce Greyson, 'Incidence of Near-Death Experiences following Attempted Suicide,' *Suicide and Life-Threatening Behavior* 16(1) (1986), 40–45; Bruce Greyson, 'Near-Death Experiences Precipitated by Suicide Attempt: Lack of Influence of Psychopathology, Religion, and Expectations,' *Journal of Near-Death Studies* 9(3) (1991), 183–88; Bruce Greyson, 'Near-Death Experiences and Anti-Suicidal Attitudes,' *Omega* 26(2) (1992), 81–89.

75. reasons for experiencers to keep their NDEs to themselves . . . A number of clinicians have explored the reluctance of some experiencers to share their stories. See, for example, Kimberly Clark, 'Clinical Interventions with Near-Death Experiencers,' in *The Near-Death Experience*, ed. by Bruce Greyson and Charles Flynn (Springfield, IL: Charles C. Thomas, 1984), 242–55; Cherie Sutherland, *Reborn in the Light* (New York: Bantam, 1995); Regina M. Hoffman, 'Disclosure Needs and Motives after a Near-Death Experience,' *Journal of Near-Death Studies* 13(4) (1995), 237–66; Regina M. Hoffman, 'Disclosure Habits after Near-Death Experiences: Influences, Obstacles, and Listener Selection,' *Journal of Near-Death Studies* 14(1) (1995), 29–48; Nancy L. Zingrone and Carlos S. Alvarado, 'Pleasurable Western Adult Near-Death Experiences: Features, Circumstances, and Incidence,' in *The Handbook of Near-Death Experiences*, ed. by Janice Miner Holden, Bruce Greyson, and Debbie James (Santa Barbara, CA: Praeger/ABC-CLIO, 2009), 17–40; L. Suzanne Gordon, 'An Ethnographic Study of Near-Death Experience Impact and Aftereffects and their Cultural Implications,' *Journal of Near-Death Studies* 31(2) (2012), 111–29; Janice Miner Holden, Lee Kinsey, and Travis R. Moore, 'Disclosing Near-Death Experiences to Professional Healthcare Providers and Nonprofessionals,' *Spirituality in Clinical Practice* 1(4) (2014), 278–87.

5. How Do We Know What's Real?
78. I developed the NDE Scale . . . The NDE Scale and its psychometric properties are described in Bruce Greyson, 'The Near-Death Experience Scale: Construction, Reliability, and Validity,' *Journal of Nervous and Mental Disease* 171(6) (1983), 369–75; Bruce Greyson, 'Near-Death Encounters with and without Near-Death Experiences: Comparative NDE Scale Profiles,' *Journal of Near-Death Studies* 8(3) (1990), 151–61; and Bruce Greyson, 'Consistency of Near-Death Experience Accounts over Two Decades: Are Reports Embellished over Time?' *Resuscitation* 73(3) (2007), 407–11.
80. a tunnel is something our minds imagine . . . Kevin Drab, 'The Tunnel Experience: Reality or Hallucination?' *Anabiosis* 1(2) (1981), 126–52; C. T. K. Chari, 'Parapsychological Reflections on Some Tunnel Experiences,' *Anabiosis* 2 (1982), 110–31.
80. compared to what theoretical physicists call a 'wormhole' . . . J. Kenneth Arnette, 'On the Mind/Body Problem: The Theory of Essence,' *Journal of Near-Death Studies* 11(1) (1992), 5–18.
82. their analysis ended up confirming the validity of the NDE Scale . . . Rense Lange, Bruce Greyson, and James Houran, 'A Rasch Scaling Validation of a "Core" Near-Death Experience,' *British Journal of Psychology* 95 (2004), 161–77.

82. Ian and I published a short article on near-death experiences ...
Ian Stevenson and Bruce Greyson, 'Near-Death Experiences: Relevance to the Question of Survival after Death,' *JAMA* 242(3) (1979), 265–67.

83. NDEs often contradict experiencers' religious beliefs about an afterlife: Henry Abramovitch, 'An Israeli Account of a Near-Death Experience: A Case Study of Cultural Dissonance,' *Journal of Near-Death Studies* 6(3) (1988), 175–84; Mark Fox, *Religion, Spirituality, and the Near-Death Experience* (London: Routledge, 2003); Kenneth Ring, *Heading Toward Omega* (New York: Coward, McCann & Geoghegan, 1984).

85. The complaining letter and our response were published together ... Monroe Schneider, 'The Question of Survival after Death,' *JAMA* 242(24) (1979), 2665; Ian Stevenson and Bruce Greyson, 'The Question of Survival after Death – Reply,' *JAMA* 242(24) (1979), 2665.

85. several additional articles in major psychiatry journals ... For example, Bruce Greyson and Ian Stevenson, 'The Phenomenology of Near-Death Experiences,' *American Journal of Psychiatry* 137(10) (1980), 1193–96; Bruce Greyson, 'Near-Death Experiences and Personal Values,' *American Journal of Psychiatry* 140(5) (1983), 618–20; and Bruce Greyson, 'The Psychodynamics of Near-Death Experiences,' *Journal of Nervous and Mental Disease* 171(6) (1983), 376–81.

89. 'The basic data for study *must* come from the stories' ... This quote appears on page 273 of Arvin S. Gibson, 'Review of Melvin Morse's *Transformed by the Light,*' *Journal of Near-Death Studies* 13(4) (1995), 273–75.

90. 'The plural of anecdote is data' ... This oft-repeated quote from Raymond Wolfinger, which he coined in the late 1960s while teaching a graduate seminar at Stanford University, was later misconstrued by debunkers as 'the plural of anecdote is not data.' It appears on page 779 of Nelson W. Polsby, 'The Contributions of President Richard F. Fenno, Jr.,' *PS: Political Science and Politics* 17(4) (1984), 778–81; and on page 83 of Nelson W. Polsby, 'Where Do You Get Your Ideas?' *PS: Political Science and Politics* 26(1) (1993), 83–87.

90. there are many fields that everyone accepts as science ... Jared Diamond, 'A New Scientific Synthesis of Human History,' in *The New Humanists*, ed. by John Brockman (New York: Barnes & Noble Books, 2003).

91. there is no evidence supporting the usefulness of parachutes ... The two quotes cited here appear on pages 1459 and 1461 of Gordon C. S. Smith and Jill P. Pell, 'Parachute Use to Prevent Death and Major Trauma Related to Gravitational Challenge: Systematic Review of Randomised Controlled Trials,' *BMJ* 327(7429) (2003), 1459–61.

6. OUT OF THEIR BODIES

94. scientific advances often come about when a new fact is discovered that can't be explained ... Thomas Kuhn, *The Structure of Scientific Revolutions* (Chicago: University of Chicago Press, 1962), chapter 6.

94. that's when Al Sullivan came into my life ... Al described his NDE in a self-published and undated booklet, *Roadway to the Lights*. His NDE was also discussed in Emily Williams Cook, Bruce Greyson, and Ian Stevenson, 'Do Any Near-Death Experiences Provide Evidence for the Survival of Human Personality after Death? Relevant Features and Illustrative Case Reports,' *Journal of Scientific Exploration* 12(3) (1998) 377–406; and in Emily Williams Kelly, Bruce Greyson, and Ian Stevenson, 'Can Experiences Near Death Furnish Evidence of Life after Death?' *Omega* 40(4) (2000), 513–19.

104. He described repeated out-of-body experiences ... Ogston described his NDE on pages 222–33 of his autobiography, Alexander Ogston, *Reminiscences of Three Campaigns* (London: Hodder and Stoughton, 1919).

105. 'I felt like a genie liberated from its bottle' ... This quote appears on page 67 of Jill Bolte Taylor, *My Stroke of Insight* (New York: Viking/Penguin, 2006).

105. Sabom found that experiencers' descriptions of their resuscitations were highly accurate ... Michael Sabom, *Recollections of Death* (New York: Harper & Row, 1982).

106. Penny Sartori replicated Sabom's findings ... Penny Sartori, *The Near-Death Experiences of Hospitalized Intensive Care Patients* (Lewiston, NY: Edwin Mellen Press, 2008).

106. ninety-three reports of out-of-body perceptions ... Janice Miner Holden, 'Veridical Perception in Near-Death Experiences,' in *The Handbook of Near-Death Experiences*, ed. by Janice Miner Holden, Bruce Greyson, and Debbie James (Santa Barbara, CA: Praeger/ABC-CLIO, 2009), 185–211.

106. 'it is enough if you prove one single crow to be white' ... This quote appears on page 5 of William James, 'Address by the President,' *Proceedings of the Society for Psychical Research* 12(1) (1897), 2–10.

107. attempts to test the accuracy of out-of-body perceptions during NDEs ... Janice Miner Holden and Leroy Joesten, 'Near-Death Veridicality Research in the Hospital Setting: Problems and Promise,' *Journal of Near-Death Studies* 9(1) (1990), 45–54; Madelaine Lawrence, *In a World of Their Own* (Westport, CT: Praeger, 1997); Sam Parnia, Derek G. Waller, Rebekah Yeates, and Peter Fenwick, 'A Qualitative and

Quantitative Study of the Incidence, Features and Aetiology of Near Death Experiences in Cardiac Arrest Survivors,' *Resuscitation* 48(2) (2001), 149–56; Penny Sartori, *The Near-Death Experiences of Hospitalized Intensive Care Patients* (Lewiston, NY, Edwin Mellen Press, 2008); Bruce Greyson, Janice Miner Holden, and J. Paul Mounsey, 'Failure to Elicit Near-Death Experiences in Induced Cardiac Arrest,' *Journal of Near-Death Studies* 25(2) (2006), 85–98; Sam Parnia, Ken Spearpoint, Peter Fenwick, et al., 'AWARE-AWAreness during REsuscitation – A Prospective Case Study,' *Resuscitation* 85(12) (2014), 1799–1805.

107. A 2018 update from the American Heart Association . . . Emilia J. Benjamin, Salim S. Virani, Clifton W. Callaway, et al., 'Heart Disease and Stroke Statistics – 2018 Update: A Report from the American Heart Association,' *Circulation* 137(12) (2018), e67–e492.

108. a study with patients who I *knew* would survive the cardiac arrest . . . Bruce Greyson, Janice Miner Holden, and J. Paul Mounsey, 'Failure to Elicit Near-Death Experiences in Induced Cardiac Arrest,' *Journal of Near-Death Studies* 25(2) (2006), 85–98.

108. Cathy Milne, a cardiac clinic nurse, had studied the frequency of NDEs . . . Catherine T. Milne, 'Cardiac Electrophysiology Studies and the Near-Death Experience,' *CACCN: The Journal of the Canadian Association of Critical Cared Nurses* 6(1) (1995), 16–19.

111. 'Anomalies tend to get swept under the carpet' . . . This quote appears on page 72 of Charles Whitehead, 'Everything I Believe Might Be a Delusion. Whoa! Tucson 2004: Ten Years On, and Are We Any Nearer to a Science of Consciousness?' *Journal of Consciousness Studies* 11(12) (2004), 68–88.

7. OR OUT OF THEIR MINDS?

116. I used the Psychiatric Diagnostic Screening Questionnaire . . . Mark Zimmerman and Jill I. Mattia, 'A Self-Report Scale to Help Make Psychiatric Diagnoses,' *Archives of General Psychiatry* 58(8) (2001), 787–94.

117. the Dissociative Experiences Scale . . . Eve Bernstein and Frank Putnam, 'Development, Reliability, and Validity of a Dissociation Scale,' *Journal of Nervous and Mental Disease* 174(12) (1986), 727–35; Bruce Greyson, 'Dissociation in People Who Have Near-Death Experiences: Out of Their Bodies or Out of Their Minds?' *Lancet* 355(9202) (2000), 460–63.

118. the Impact of Event Scale . . . Mardi Horowitz, Nancy Wilner, and William Alvarez, 'Impact of Event Scale: A Measure of Subjective Stress,'

Psychosomatic Medicine 41(3) (1979), 209–18; Bruce Greyson, 'Posttraumatic Stress Symptoms following Near-Death Experiences,' *American Journal of Orthopsychiatry* 71(3) (2001), 368–73.

120. the revised ninety-item Symptom Checklist . . . Leonard Derogatis, *SCL-90-R Administration, Scoring, and Procedures Manual – II* (Towson, MD: Clinical Psychometric Research, 1992); Bruce Greyson, 'Near-Death Experiences in a Psychiatric Outpatient Clinic Population,' *Psychiatric Services* 54(12) (2003), 1649–51.

125. a twenty-five-year-old nurse who had an NDE . . . This case was described on page 71 of Bruce Greyson, 'Is Consciousness Produced by the Brain?' in *Cosmology and Consciousness*, ed. by Bryce Johnson (Dharamsala, India: Library of Tibetan Works and Archives, 2013), 59–87.

128. Mitch Liester and I compared two groups of people . . . Bruce Greyson and Mitchell Liester, 'Auditory Hallucinations Following Near-Death Experiences,' *Journal of Humanistic Psychology* 44(3) (2004), 320–36.

129. unusual experiences that are symptoms of mental illness and those like NDEs . . . Bruce Greyson, 'Differentiating Spiritual and Psychotic Experiences: Sometimes a Cigar Is Just a Cigar,' *Journal of Near-Death Studies* 32(3) (2014), 123–36. I drew on the work of a number of scholars to delineate this distinction, including Janice Miner Holden, in *Near-Death Experiences*, produced by Roberta Moore (Fort Myers, FL: Blue Marble Films, 2013); Harold G. Koenig, 'Religion, Spirituality, and Psychotic Disorders,' *Revista de Psiquiatria Clinica* 34 (Supplement 1) (2007), 40–48; David Lukoff, 'Visionary Spiritual Experiences,' *Southern Medical Journal* 100(6) (2007), 635–41; Penny Sartori, 'A Prospective Study of NDEs in an Intensive Therapy Unit,' *Christian Parapsychologist* 16(2) (2004), 34–40; Penny Sartori, *The Near-Death Experiences of Hospitalized Intensive Care Patients* (Lewiston, NY: Edwin Mellen Press, 2008); Adair Menezes and Alexander Moreira-Almeida, 'Differential Diagnosis between Spiritual Experiences and Mental Disorders of Religious Content,' *Revista de Psiquiatria Clinica*, 36(2) (2009), 75–82; Adair Menezes and Alexander Moreira-Almeida, 'Religion, Spirituality, and Psychosis,' *Current Psychiatry Reports* 12(3) (2010), 174–79; Alexander Moreira-Almeida, 'Assessing Clinical Implications of Spiritual Experiences,' *Asian Journal of Psychiatry* 5(4) (2012), 344–46; Alexander Moreira-Almeida and Etzel Cardeña, 'Differential Diagnosis between Non-Pathological Psychotic and Spiritual Experiences and Mental

Disorders: A Contribution from Latin American Studies to the ICD-11,' *Revista Brasileira de Psiquiatria* 33 (Supplement 1) (2011), 529–89; and Kathleen D. Noble, 'Psychological Health and the Experience of Transcendence,' *Counseling Psychologist* 15(4) (1984), 601–14.

129. Their memory doesn't fade over time . . . Bruce Greyson, 'Consistency of Near-Death Experience Accounts over Two Decades: Are Reports Embellished over Time?' *Resuscitation* 73(3) (2007), 407–11; Lauren E. Moore and Bruce Greyson, 'Characteristics of Memories for Near-Death Experiences,' *Consciousness and Cognition* 51 (2017), 116–24.

130. some people do learn and grow from their mental illnesses . . . Gary Nixon, Brad Hagen, and Tracey Peters, 'Psychosis and Transformation: A Phenomenological Inquiry,' *International Journal of Mental Health and Addiction* 8(4) (2010), 527–44.

8. ARE NEAR-DEATH EXPERIENCES REAL?

132. 'Science is not defined by the topics it studies but rather by its approach to investigating those topics' . . . This quote appears on pages 275–76 in Mark Leary, 'Why Are (Some) Scientists so Opposed to Parapsychology?' *Explore* 7(5) (2011), 275–77.

133. scientists thought reports of meteorites were tall tales . . . Kat Eschner, 'Scientists Didn't Believe in Meteorites until 1803,' *Smithsonian Magazine*, April 26, 2017, www.smithsonianmag.com/smart-news/ 1803-rain-rocks-helped-establish-existence-meteorites-180963017/.

133. physicians ridiculed the idea of germs . . . John Waller, *The Discovery of the Germ* (New York: Columbia University Press, 2003).

133. bacteria that might cause stomach ulcers . . . Richard B. Hornick, 'Peptic Ulcer Disease: A Bacterial Infection?' *New England Journal of Medicine* 316(25) (1987), 1598–1600.

134. Refining our models when new phenomena are discovered . . . Lisa Feldman Barrett, 'Psychology Is Not in Crisis,' *New York Times*, September 1, 2015, www.nytimes.com/2015/09/01/opinion/psychology-is -not-in-crisis.html/.

135. 'Science is not about finding the truth at all' . . . Thomas M. Schofield, 'On My Way to Being a Scientist,' *Nature* 497 (2013), 277–78.

135. 'Objective truth,' Tyson says, 'is the kind of truth science discovers' . . . Neil deGrasse Tyson, 'Neil deGrasse Tyson on Death and Near Death Experiences,' excerpt from lecture, May 3, 2017, 92nd Street Y, www.youtube.com/watch?v=y5qEBC7ZzVQ.

136. a strange and wondrous beast: a camel . . . Kaplan's story is related on page 379 of Paul C. Horton, 'The Mystical Experience: Substance of an Illusion,' *Journal of the American Psychoanalytic Association* 22(2) (1974), 364–80.

137. if you get hit by a truck, you *know* that you were hit by a truck . . . This quote appears on page 7 of Robert L. Van de Castle, 'The Concept of Porosity in Dreams,' *EdgeScience* 14 (2013), 6–10.

138. there wouldn't be counterfeit gold unless there was real gold . . . This quote from Rumi was cited by Idries Shah in Elizabeth Hall, 'The Sufi Tradition: Interview with Idries Shah,' *Psychology Today*, July 1975, www.katinkahesselink.net/sufi/sufi-shah.html.

138. Many NDEs occur in cardiac arrest, which often causes amnesia . . . Sam Parnia, Ken Spearpoint, and Peter B. Fenwick, 'Near Death Experiences, Cognitive Function and Psychological Outcomes of Surviving Cardiac Arrest,' *Resuscitation* 74(2) (2007), 215–21.

138. NDEs sometimes happen to people who have taken psychedelic drugs, which can interfere with memory . . . H. Valerie Curran, 'Psychopharmacological Perspectives on Memory,' in *The Oxford Handbook of Memory*, ed. by Endel Tulving and Fergus Craik (New York: Oxford University Press, 2000), 539–54.

138. NDEs usually occur in traumatic situations, which are known to influence the accuracy of memories . . . Jonathan W. Schooler and Eric Eich, 'Memory for Emotional Events,' in *The Oxford Handbook of Memory*, ed. by Endel Tulving and Fergus Craik (New York: Oxford University Press, 2000), 379–92.

138. they usually include strong positive emotions, which may influence memory . . . Alexandre Schaefer and Pierre Philippot, 'Selective Effects of Emotion on the Phenomenal Characteristics of Autobiographical Memories,' *Memory* 13(2) (2005), 148–60.

138. sometimes reported long after the event, which often reduces the detail and vividness of memories . . . Lucia M. Talamini and Eva Goree, 'Aging Memories: Differential Decay of Episodic Memory Components,' *Learning and Memory* 19(6) (2012), 239–46.

138. researchers have speculated that NDE stories are embellished over time . . . Nathan Schnaper, 'Comments Germane to the Paper Entitled "The Reality of Death Experiences" by Ernst Rodin,' *Journal of Nervous and Mental Disease* 168(5) (1980), 268–70.

138. What I found was that the accounts had *not* become more blissful over time . . . Bruce Greyson, 'Consistency of Near-Death

Experience Accounts over Two Decades: Are Reports Embellished over Time?' *Resuscitation* 73(3) (2007), 407–11.

139. I selected two dozen of Ian's most complete cases . . . Geena Athappilly, Bruce Greyson, and Ian Stevenson, 'Do Prevailing Societal Models Influence Reports of Near-Death Experiences? A Comparison of Accounts Reported before and after 1975,' *Journal of Nervous and Mental Disease* 194(3) (2006), 218–22.

140. most experiencers are quite certain about the reality of their NDEs . . . Andrew J. Dell'Olio, 'Do Near-Death Experiences Provide a Rational Basis for Belief in Life after Death?' *Sophia* 49(1) (2010), 113–28.

140. Jeffrey Long found that 96 percent rated their NDEs as 'definitely real' . . . Jeffrey Long (with Paul Perry), *Evidence of the Afterlife* (New York: HarperOne, 2010).

141. LeaAnn Carroll developed a massive blood clot in her lung . . . LeaAnn's medical crisis was described in Alan T. Marty, Frank L. Hilton, Robert K. Spear, and Bruce Greyson, 'Postcesarean Pulmonary Embolism, Sustained Cardiopulmonary Resuscitation, Embolectomy, and Near-Death Experience,' *Obstetrics and Gynecology* 106(5 Pt. 2) (2005), 1153–55. She described her NDE in LeaAnn Carroll, *There Stood a Lamb* (Kearney, NE: Morris Publications, 2004).

141. Nancy Evans Bush, who had an NDE at age twenty-seven . . . Nancy described her NDE in Nancy Evans Bush, *Dancing Past the Dark* (Cleveland, TN: Parson's Porch Books, 2012).

141. I used the Memory Characteristics Questionnaire . . . Lauren E. Moore and Bruce Greyson, 'Characteristics of Memories for Near-Death Experiences,' *Consciousness and Cognition* 51 (2017), 116–24.

142. Two other research teams, in Belgium and in Italy, came up with the same results . . . Charlotte Martial, Vanessa Charland-Verville, Héléna Cassol, et al., 'Intensity and Memory Characteristics of Near-Death Experiences,' *Consciousness and Cognition* 56 (2017), 120–27; Arianna Palmieri, Vincenzo Calvo, Johann R. Kleinbub, et al., '"Reality" of Near-Death Experience Memories: Evidence from a Psychodynamic and Electrophysiological Integrated Study,' *Frontiers in Human Neuroscience* 8 (2014), 429.

9. THE BIOLOGY OF DYING

144. researchers have argued for NDEs being associated with the right temporal lobe . . . Olaf Blanke, Stéphanie Ortigue, Theodor Landis, and Margitta Seeck, 'Stimulating Illusory Own-Body Perceptions,' *Nature* 419(6904) (2002), 269–70.

144. others argued for the left . . . Willoughby B. Britton and Richard R. Bootzin, 'Near-Death Experiences and the Temporal Lobe,' *Psychological Science* 15(4) (2004), 254–58.

145. Penfield is widely believed to have produced out-of-body experiences . . . Susan Blackmore, *Dying to Live* (Amherst, NY: Prometheus, 1993); Melvin L. Morse, David Venecia, and Jerrold Milstein, 'Near-Death Experiences: A Neurophysiological Explanatory Model,' *Journal of Near-Death Studies* 8(1) (1989), 45–53; Vernon M. Neppe, 'Near-Death Experiences: A New Challenge in Temporal Lobe Phenomenology? Comments on "A Neurobiological Model for Near-Death Experiences,"' *Journal of Near-Death Studies* 7(4) (1989), 243–48; Frank Tong, 'Out-of-Body Experiences: From Penfield to Present,' *Trends in Cognitive Science* 7(3) (2003), 104–6.

146. 'Oh, God! I am leaving my body' . . . This quote appears on page 458 of Wilder Penfield, 'The Twenty-Ninth Maudsley Lecture: The Role of the Temporal Cortex in Certain Psychical Phenomena,' *Journal of Mental Science* 101(424) (1955), 451–65.

146. 'I have a queer sensation as if I am not here' . . . This quote and the next four in this paragraph were reported on page 174 of Wilder Penfield and Theodore Rasmussen, *The Cerebral Cortex of Man* (New York: Macmillan, 1950).

146. patients with a variety of seizure types affecting different parts of the brain . . . Orrin Devinsky, Edward Feldmann, Kelly Burrowes, and Edward Bromfield, 'Autoscopic Phenomena with Seizures,' *Archives of Neurology* 46(10) (1989), 1080–88.

147. NDEs and similar experiences are associated with different parts of the brain . . . Nina Azari, Janpeter Nickel, Gilbert Wunderlich, et al., 'Neural Correlates of Religious Experience,' *European Journal of Neuroscience* 13(8) (2001), 1649–52; Peter Fenwick, 'The Neurophysiology of Religious Experience,' in *Psychosis and Spirituality*, ed. by Isabel Clarke (London: Whurr, 2001), 15–26; Andrew B. Newberg and Eugene G. d'Aquili, 'The Near Death Experience as Archetype: A Model for "Prepared" Neurocognitive Processes,' *Anthropology of Consciousness* 5(4) (1994), 1–15.

147. brain activity in people who had had NDEs . . . Mario Beauregard, Jérôme Courtemanche, and Vincent Paquette, 'Brain Activity in Near-Death Experiencers During a Meditative State,' *Resuscitation* 80(9) (2009), 1006–10.

151. I then interviewed a hundred patients in the epilepsy clinic . . .
Bruce Greyson, Nathan B. Fountain, Lori L. Derr, and Donna
K. Broshek, 'Out-of-Body Experiences Associated with Seizures,' *Frontiers
in Human Neuroscience* 8(65) (2014), 1–11; Bruce Greyson, Nathan
B. Fountain, Lori L. Derr, and Donna K. Broshek, 'Mystical Experiences
Associated with Seizures,' *Religion, Brain & Behavior* 5(3) (2015),
182–96.

**154. patients who described a sense of leaving their bodies during
seizures commonly report intense horror or fear . . .** Peter Brugger,
Reto Agosti, Marianne Regard, et al., 'Heautoscopy, Epilepsy, and
Suicide,' *Journal of Neurology, Neurosurgery, and Psychiatry* 57(7) (1994),
838–39; Devinsky et al., 'Autoscopic Phenomena with Seizures.'

**155. 'I was conscious that my mental self used regularly to leave the
body' . . .** This quote appears on page 222 of Alexander Ogston,
Reminiscences of Three Campaigns (London: Hodder and Stoughton,
1919).

156. 'I felt like a genie liberated from its bottle' . . . This quote appears
on page 67 of Jill Bolte Taylor, *My Stroke of Insight* (New York: Viking/
Penguin, 2006).

**158. stimulating the temporal lobe of a patient with an electric
current can produce a feeling of the body being distorted . . .** See, for
example, Olaf Blanke, Stéphanie Ortigue, Theodor Landis, and Margitta
Seeck, 'Stimulating Illusory Own-Body Perceptions,' *Nature* 419(6904)
(2002), 269–70.

**158. there are many important differences between these
sensations . . .** Bruce Greyson, Sam Parnia, and Peter Fenwick,
'[Comment on] Visualizing Out-of-Body Experience in the Brain,' *New
England Journal of Medicine* 358(8) (2008), 855–56.

**159. brain activity decreases within six to seven seconds of the heart
stopping . . .** Pim van Lommel, 'Near-Death Experiences: The Experience
of the Self as Real and Not as an Illusion,' *Annals of the New York Academy
of Sciences* 1234(1) (2011), 19–28; Jaap W. de Vries, Patricia F. A. Bakker,
Gerhard H. Visser, et al., 'Changes in Cerebral Oxygen Uptake and
Cerebral Electrical Activity during Defibrillation Threshold Testing,'
Anesthesia and Analgesia 87(1) (1998), 16–20; Holly L. Clute and Warren
J. Levy, 'Electroencephalographic Changes during Brief Cardiac Arrest in
Humans,' *Anesthesiology* 73 (1990), 821–25; Thomas J. Losasso, Donald
A. Muzzi, Frederic B. Meyer, and Frank W. Sharbrough,
'Electroencephalographic Monitoring of Cerebral Function during

Asystole and Successful Cardiopulmonary Resuscitation,' *Anesthesia and Analgesia* 75(6) (1992), 1021–24.

159. after the heart stops there is *no* well-defined EEG activity . . . Loretta Norton, Raechelle M. Gibson, Teneille Gofton, et al., 'Electroencephalographic Recordings during Withdrawal of Life-Sustaining Therapy until 30 Minutes after Declaration of Death,' *Canadian Journal of Neurological Sciences* 44(2) (2017), 139–45.

160. rapid eye movement (REM) brain activity, can intrude into our waking thoughts . . . Kevin R. Nelson, Michelle Mattingly, Sherman A. Lee, and Frederick A. Schmitt, 'Does the Arousal System Contribute to Near Death Experience?' *Neurology* 66(7) (2006), 1003–9.

160. no higher than the rate of the same REM intrusion symptoms in a random sample . . . Bruce Greyson and Jeffrey P. Long, '[Comment on] Does the Arousal System Contribute to Near Death Experience?' *Neurology* 67(12) (2006), 2265; Maurice M. Ohayon, Robert G. Priest, Jürgen Zully, et al., 'Prevalence of Narcolepsy Symptomatology and Diagnosis in the European General Population' *Neurology* 58(12) (2002), 1826–33.

160. people under general anesthesia, which suppresses REM brain activity . . . Arthur J. Cronin, John Keifer, Matthew F. Davies, et al., 'Postoperative Sleep Disturbance: Influences of Opioids and Pain in Humans,' *Sleep* 24(1) (2001), 39–44.

160. measurements of REM brain activity in people who've had NDEs show that it was actually *lower* . . . Britton and Bootzin, 'Near-Death Experiences and the Temporal Lobe.'

160. experiencers remembering their NDEs did not have brain wave patterns typical of recalling fantasies . . . Arianna Palmieri, Vincenzo Calvo, Johann R. Kleinbub, et al., ' "Reality" of Near-Death Experience Memories: Evidence from a Psychodynamic and Electrophysiological Integrated Study,' *Frontiers in Human Neuroscience* 8 (2014), 429.

160. decreased oxygen in the brain might be a factor in causing NDEs . . . See, for example, James E. Whinnery, 'Psychophysiologic Correlates of Unconsciousness and Near-Death Experiences,' *Journal of Near-Death Studies* 15(4) (1997), 231–58.

161. decreased oxygen is a very unpleasant experience . . . William Breitbart, Christopher Gibson, and Annie Tremblay, 'The Delirium Experience: Delirium Recall and Delirium-Related Distress in Hospitalized Patients with Cancer, Their Spouses/Caregivers, and Their Nurses,' *Psychosomatics* 43(3) (2002), 183–94.

161. very different from NDEs, which are usually peaceful, positive experiences . . . Nancy L. Zingrone and Carlos S. Alvarado, 'Pleasurable Western Adult Near-Death Experiences: Features, Circumstances, and Incidence,' in *The Handbook of Near-Death Experiences*, ed. by Janice Miner Holden, Bruce Greyson, and Debbie James (Santa Barbara, CA: Praeger/ABC-CLIO, 2009), 17–40.

161. NDEs are associated either with *increased* oxygen levels . . . Sam Parnia, Derek G. Waller, Rebekah Yeates, and Peter Fenwick, 'A Qualitative and Quantitative Study of the Incidence, Features and Aetiology of Near Death Experiences in Cardiac Arrest Survivors,' *Resuscitation* 48(2) (2001), 149–56; Michael Sabom, *Recollections of Death* (New York: Harper & Row, 1982).

161. or with levels the same as those of non-experiencers . . . Melvin Morse, Doug Conner, and Donald Tyler, 'Near-Death Experiences in a Pediatric Population: A Preliminary Report,' *American Journal of Diseases of Children* 139(6) (1985), 595–600; Pim van Lommel, Ruud van Wees, Vincent Meyers, and Ingrid Elfferich, 'Near-Death Experiences in Survivors of Cardiac Arrest: A Prospective Study in the Netherlands,' *Lancet* 358(9298) (2001), 2039–45.

161. patients who are given medications in fact report *fewer* NDEs . . . Bruce Greyson, 'Organic Brain Dysfunction and Near-Death Experiences,' paper presented at the American Psychiatric Association 135th Annual Meeting, Toronto, May 15–21, 1982; Karlis Osis and Erlendur Haraldsson, *At the Hour of Death* (New York: Avon, 1977); Sabom, *Recollections of Death*.

161. most often those associated with the anesthetic ketamine . . . Karl L. R. Jansen, 'The Ketamine Model of the Near-Death Experience: A Central Role for the N-Methyl-D-Aspartate Receptor,' *Journal of Near-Death Studies* 16(1) (1997), 5–26; Ornella Corazza and Fabrizio Schifano, 'Near-Death States Reported in a Sample of 50 Misusers,' *Substance Use and Misuse* 45(6) (2010), 916–24.

161. and with DMT (dimethyltryptamine) . . . Rick Strassman, *DMT* (Rochester, VT: Park Street Press, 2001); Christopher Timmermann, Leor Roseman, Luke Williams, et al., 'DMT Models the Near-Death Experience,' *Frontiers in Psychology* 9 (2018), 1424.

162. I was recently part of a multinational research team that analyzed language . . . Charlotte Martial, Héléna Cassol, Vanessa Charland-Verville, et al., 'Neurochemical Models of Near-Death Experiences: A Large-Scale Study Based on the Semantic Similarity of Written Reports,' *Consciousness and Cognition* 69 (2019), 52–69.

162. he viewed ketamine as 'just another door' to NDEs . . . Karl L. R.
Jansen, 'Response to Commentaries on "The Ketamine Model of the
Near-Death Experience . . . ," ' *Journal of Near-Death Studies* 16(1)
(1997), 79–95.

162. most likely to be associated with NDEs were *endorphins* . . .
Daniel Carr, 'Pathophysiology of Stress-Induced Limbic Lobe
Dysfunction: A Hypothesis for NDEs,' *Anabiosis* 2(1) (1982), 75–89.

**162. NDEs might be connected to serotonin, adrenaline, vasopressin,
and glutamate** . . . Jansen, 'The Ketamine Model of the Near-Death
Experience'; Melvin L. Morse, David Venecia, and Jerrold Milstein,
'Near-Death Experiences: A Neurophysiologic Explanatory Model,'
Journal of Near-Death Studies 8(1) (1989), 45–53; Juan C. Saavedra-
Aguilar, and Juan S. Gómez-Jeria, 'A Neurobiological Model for
Near-Death Experiences,' *Journal of Near-Death Studies* 7(4) (1989),
205–22.

163. the blind men and the elephant . . . John Ireland, translator, *The
Udāna and the Itivuttaka* (Kandy, Sri Lanka: Buddhist Publication Society,
2007).

10. THE BRAIN AT DEATH

167. the experience of neurosurgeon Eben Alexander . . . Eben
described his near-death experience in Eben Alexander, *Proof of Heaven*
(New York: Simon & Schuster, 2012).

**169. All three of us concluded independently that he had been
extremely close to death** . . . Surbhi Khanna, Lauren E. Moore, and
Bruce Greyson, 'Full Neurological Recovery from *Escherichia coli*
Meningitis Associated with Near-Death Experience,' *Journal of Nervous
and Mental Disease* 206(9) (2018), 744–47.

170. 'The mind is what the brain does' . . . This quote appears on page
4 of Stephen M. Kosslyn and Olivier M. Koenig, *Wet Mind* (New York:
Free Press/Macmillan, 1992).

172. 'I didn't make that sound. You pulled it out of me' . . . This quote
appears on pages 76–77 of Wilder Penfield, *Mystery of the Mind*
(Princeton, NJ: Princeton University Press, 1975).

173. 'we don't have a clue' . . . This quote appears on page xi of Alva
Noë, *Out of Our Heads* (New York: Hill and Wang, 2009).

174. 'it has something to do with the head, rather than the foot' . . .
This quote appears on page 249 of Nick Herbert, *Quantum Reality*
(Garden City, NY: Anchor/Doubleday, 1985).

174. 'as fantastic as the idea of a self-playing orchestra' . . . This quote appears on page 64 of Noë, *Out of Our Heads*.

175. the mind being a function of the brain can be interpeted in two very different ways . . . William James, *Human Immortality* (Boston: Houghton Mifflin, 1898).

11. THE MIND IS NOT THE BRAIN

180. 'Imagine, if you will, a huge, dark warehouse' . . . This quote appears on pages 71–73 of Anita Moorjani, *Dying to Be Me* (Carlsbad, CA: Hay House, 2014).

182. I participated in a symposium at the United Nations on alternative models . . . 'Beyond the Mind-Body Problem: New Paradigms in the Science of Consciousness,' September 11, 2008, New York, www.nourfoundation.com/events/Beyond-the-Mind-Body-Problem-New-Paradigms-in-the-Science-of-Consciousness.html

183. the majority believed the mind and brain are two separate things . . . Athena Demertzi, Charlene Liew, Didier Ledoux, et al., 'Dualism Persists in the Science of Mind,' *Annals of the New York Academy of Sciences* 1157(1) (2009), 1–9.

183. the majority believed the mind was independent of the brain . . . Alexander Moreira-Almeida and Saulo de Freitas Araujo, 'Does the Brain Produce the Mind? A Survey of Psychiatrists' Opinions,' *Archives of Clinical Psychiatry* 42(3) (2015), 74–75.

183. 'the theory needs a second look' . . . This quote appears on pages 1104–5 of Basil A. Eldadah, Elena M. Fazio, and Kristina A. McLinden, 'Lucidity in Dementia: A Perspective from the NIA,' *Alzheimer's & Dementia* 15(8) (2019), 1104–6.

184. 'it is the organ of attention to life' . . . This quote appears on page 168 of Henri Bergson, 'Presidential Address' (translated by H. Wildon Carr), *Proceedings of the Society for Psychical Research* 27(68) (1914), 157–75.

188. described over the centuries with various metaphors . . . Michael Grosso, 'The "Transmission" Model of Mind and Body: A Brief History,' in *Beyond Physicalism*, ed. by Edward F. Kelly, Adam Crabtree, and Paul Marshall (Lanham, MD: Rowman & Littlefield, 2015), 79–113.

188. 'the brain is the interpreter of consciousness' . . . This quote appears on page 179 of Hippocrates, *Hippocrates. Volume 2: The Sacred Disease, Sections XIX & XX*, translated by William Henry Samuel Jones (Cambridge, MA: Harvard University Press/Loeb Classical Library, 1923). (Original work written around 400 BC.)

188. 'funneled through the reducing valve of the brain' . . . This quote appears on pages 22–24 of Aldous Huxley, *The Doors of Perception* (New York: Perennial Library/Harper & Row, 1954).

190. biological mechanisms by which the brain could act as a filter . . . Edward F. Kelly and David E. Presti, 'A Psychobiological Perspective on "Transmission" Models,' in *Beyond Physicalism*, ed. by Edward F. Kelly, Adam Crabtree, and Paul Marshall (Lanham, MD: Rowman & Littlefield, 2015), 115–55; Marjorie Woollacott and Anne Shumway-Cook, 'The Mystical Experience and Its Neural Correlates,' *Journal of Near-Death Studies*, 38 (2020), 3–25.

190. A similar unexplained experience is something called 'terminal lucidity' . . . Michael Nahm, Bruce Greyson, Emily W. Kelly, and Erlendur Haraldsson, 'Terminal Lucidity: A Review and a Case Collection,' *Archives of Gerontology and Geriatrics* 55(1) (2012), 138–42.

190. a workshop at the National Institute on Aging . . . George A. Mashour, Lori Frank, Alexander Batthyany, et al., 'Paradoxical Lucidity: A Potential Paradigm Shift for the Neurobiology and Treatment of Severe Dementias,' *Alzheimer's & Dementia* 15(8) (2019), 1107–14.

191. neuroimaging studies of people under the influence of psychedelic drugs . . . Robin L. Carhart-Harris, David Erritzoe, Tim Williams, et al., 'Neural Correlates of the Psychedelic State as Determined by fMRI Studies with Psilocybin,' *Proceedings of the National Academy of Sciences* 109(6) (2012), 2138–43; Robin L. Carhart-Harris, Suresh D. Muthukumaraswamy, Leor Roseman, et al., 'Neural Correlates of the LSD Experience Revealed by Multimodal Neuroimaging,' *Proceedings of the National Academy of Sciences* 113(17) (2016), 4853–58; Suresh D. Muthukumaraswamy, Robin L. Carhart-Harris, Rosalyn J. Moran, et al., 'Broadband Cortical Desynchronization Underlies the Human Psychedelic State,' *Journal of Neuroscience* 33(38) (2013), 15171–83; Fernanda Palhano-Fontes, Katia Andrade, Luis Tofoli, et al., 'The Psychedelic State Induced by Ayahuasca Modulates the Activity and Connectivity of the Default Mode Network,' *PLOS ONE* 10(2) (2015), e0118143.

191. 'We are conscious not *because* of the brain but *in spite of* it' . . . This quote appears on page 191 of Larry Dossey, *The Power of Premonitions* (New York: Dutton, 2009).

12. DOES CONSCIOUSNESS CONTINUE?

194. recently deceased people *who were not known to have died* . . . Bruce Greyson, 'Seeing Deceased Persons Not Known to Have Died:

"Peak in Darien" Experiences,' *Anthropology and Humanism* 35(2) (2010), 159–71.

197. Barbara Langer told me about a similar NDE . . . An abbreviated account of Barbara's NDE appears on pages 125–27 of Julia Dreyer Brigden, *Girl: An Untethered Life* (Santa Rosa, CA: Julia Dreyer Brigden, 2019).

201. 7 percent involved seeing someone . . . Emily W. Kelly, 'Near-Death Experiences with Reports of Meeting Deceased People,' *Death Studies* 25(3) (2001), 229–49.

201. a nobleman named Corfidius . . . The story of Corfidius appears on pages 624–25 of Pliny the Elder, *Natural History, Volume 2, Books 3–7*, translated by Horace Rackham (Cambridge, MA: Harvard University Press, 1942). (Original work written AD 77.)

202. Eleanor Sidgwick wrote of an Englishwoman . . . This account appears on pages 92–93 of Eleanor M. Sidgwick, 'Notes on the Evidence, Collected by the Society, for Phantasms of the Dead,' *Proceedings of the Society for Psychical Research* 3 (1885), 69–150.

203. Dr K. M. Dale reported the case of nine-year-old Eddie . . . This account appears on pages 42–46 of Brad Steiger and Sherry Hansen Steiger, *Children of the Light* (New York: Signet, 1995).

13. Heaven or Hell?

207. Dottie Bush also described visiting a place she identified as heaven . . . Dottie's spiritual growth as a result of her near-death experience is described on pages 77 and 105 of P. M. H. Atwater, *Coming Back to Life* (New York: Dodd, Mead, 1988).

210. Nancy Evans Bush and I had collected enough accounts of distressing NDEs . . . Bruce Greyson and Nancy Evans Bush, 'Distressing Near-Death experiences,' *Psychiatry* 55(1) (1992), 95–110.

211. revered mystics like the sixteenth-century Saint Teresa of Ávila . . . Saint Teresa of Ávila, *Interior Castle* (New York: Benziger Brothers, 1912). (Original work written 1577.)

211. Saint John of the Cross . . . Saint John of the Cross, *Dark Night of the Soul* (London: John M. Watkins, 1905). (Original work written 1584.)

211. Mother Teresa of Calcutta . . . Mother Teresa, *Come Be My Light* (New York: Doubleday, 2007).

211. a message for the experiencers to turn their lives around . . . Nancy Evans Bush and Bruce Greyson, 'Distressing Near-Death Experiences: The Basics,' *Missouri Medicine* 111(6) (2014), 486–91.

212. Kat Dunkle described her hellish experience . . . Kat described her NDE in Kat Dunkle, *Falling into Darkness* (Maitland, FL: Xulon Press, 2007).

218. Róisín Fitzpatrick suffered a brain hemorrhage . . . Róisín described her NDE in Róisín Fitzpatrick, *Taking Heaven Lightly* (Dublin: Hatchette Books Ireland, 2016).

219. feeling bliss when she was stricken by an unknown illness . . . Margot described her NDE in Margot Grey, *Return from Death* (London: Arkana, 1985).

14. WHAT ABOUT GOD?

228. Kim Clark Sharp, who had an NDE when she collapsed without a pulse . . . Kim described her NDE in Kimberly Clark Sharp, *After the Light* (New York, William Morrow, 1995).

236. 'it is a good plan to have occasional doubts about one's scepticism' . . . This quote appears on page 53 of Sigmund Freud, 'New Introductory Lectures on Psycho-Analysis. Lecture XXX. Dreams and Occultism,' in *The Standard Edition of the Complete Psychological Works of Sigmund Freud, Vol. 12*, translated by James Strachey (London: Hogarth Press, 1933), 31–56.

237. the analogy of a wave in the ocean . . . Novelist Katherine Anne Porter had an NDE during a nearly fatal bout with influenza during the 1918 epidemic. In her story, 'Pale Horse, Pale Rider,' she described mingling with deceased loved ones in a heavenly environment as moving 'as a wave among waves.' See Steve Straight, 'A Wave among Waves: Katherine Anne Porter's Near-Death Experience,' *Anabiosis* 4(2) (1984), 107–23.

15. THIS CHANGES EVERYTHING

241. markedly lower anxiety about death in people like John . . . Bruce Greyson, 'Reduced Death Threat in Near-Death Experiencers,' *Death Studies* 16(6) (1992), 523–36; Russell Noyes, 'Attitude Change following Near-Death Experiences,' *Psychiatry* 43(3) (1980), 234–42; Kenneth Ring, *Heading toward Omega* (New York: Coward, McCann & Geoghegan, 1984); Michael Sabom, *Recollections of Death* (New York: Harper & Row, 1982); Charles Flynn, *After the Beyond* (Englewood Cliffs, NJ: Prentice Hall, 1986).

243. features of near-death experiences are associated with changes in death attitudes . . . Marieta Pehlivanova and Bruce Greyson, 'Which

Near-Death Experience Features Are Associated with Reduced Fear of Death?' presented at the 2019 International Association of Near-Death Studies Conference, Valley Forge, PA, August 31, 2019.

244. an experience of being free from the body would reduce people's fear of death . . . Natasha A. Tassell-Matamua and Nicole Lindsay, ' "I'm Not Afraid to Die": The Loss of the Fear of Death after a Near-Death Experience,' *Mortality* 21(1) (2016), 71–87.

246. one-fourth of them have an NDE in the course of the attempt . . . Bruce Greyson, 'Incidence of Near-Death Experiences following Attempted Suicide,' *Suicide and Life-Threatening Behavior* 16(1) (1986), 40–45.

247. Those who do have NDEs are less suicidal after the event . . . Bruce Greyson, 'Near-Death Experiences and Attempted Suicide,' *Suicide and Life-Threatening Behavior* 11(1) (1981), 10–16; Kenneth Ring and Stephen Franklin, 'Do Suicide Survivors Report Near-Death Experiences?' *Omega* 12(3) (1982), 191–208.

247. They gave me a wide variety of explanations . . . Bruce Greyson, 'Near-Death Experiences and Anti-Suicidal Attitudes,' *Omega* 26 (1992), 81–89.

252. researchers have found consistent changes in experiencers' perception . . . Russell Noyes, Peter Fenwick, Janice Miner Holden, and Sandra Rozan Christian, 'Aftereffects of Pleasurable Western Adult Near-Death Experiences,' in *The Handbook of Near-Death Experiences*, ed. by Janice Miner Holden, Bruce Greyson, and Debbie James (Santa Barbara, CA: Praeger/ABC-CLIO, 2009), 41–62; Sabom, *Recollections of Death*; Bruce Greyson, 'Near Death Experiences and Personal Values,' *American Journal of Psychiatry* 140(5) (1983), 618–20; Flynn, *After the Beyond*; Margot Grey, *Return from Death* (London: Arkana, 1985).

252. beyond what we see in people who have come close to death but didn't have NDEs . . . Ring, *Heading toward Omega*; Cherie Sutherland, *Transformed by the Light* (New York: Bantam Books, 1992); Peter Fenwick and Elizabeth Fenwick, *The Truth in the Light* (New York: Berkley Books, 1995); Zalika Klemenc-Ketis, 'Life Changes in Patients after Out-of-Hospital Cardiac Arrest,' *International Journal of Behavioral Medicine* 20(1) (2013), 7–12.

253. those who don't have NDEs often become more anxious . . . Esther M. Wachelder, Véronique R. Moulaert, Caroline van Heugten, et al., 'Life after Survival: Long-Term Daily Functioning and Quality of Life after an Out-of-Hospital Cardiac Arrest,' *Resuscitation* 80(5) (2009), 517–22.

253. the Life Changes Inventory . . . The history and development of this scale, which Kenneth Ring introduced in a preliminary version in 1980, was described in Bruce Greyson and Kenneth Ring, 'The Life Changes Inventory – Revised,' *Journal of Near-Death Studies* 23(1) (2004), 41–54.

254. 'he became one of the best conducted sailors in the ship' . . . This quote appears on page 184 of Sir Benjamin Collins Brodie, *The Works of Sir Benjamin Collins Brodie* (London: Longman, Green, Longman, Roberts, and Green, 1865).

16. WHAT DOES IT ALL MEAN?

259. aspect of their personal lives that includes something beyond the usual senses . . . Lynn G. Underwood, 'Ordinary Spiritual Experience: Qualitative Research, Interpretive Guidelines, and Population Distribution for the Daily Spiritual Experience Scale,' *Archive for the Psychology of Religion* 28(1) (2006), 181–218.

260. a personal search for inspiration, meaning, and purpose . . . Eltica de Jager Meezenbroek, Bert Garssen, Machteld van den Berg, et al., 'Measuring Spirituality as a Universal Human Experience: A Review of Spirituality Questionnaires,' *Journal of Religion and Health* 51(2) (2012), 336–54.

260. those who had NDEs were significantly more satisfied with life . . . Bruce Greyson, 'Near-Death Experiences and Satisfaction with Life,' *Journal of Near-Death Studies* 13(2) (1994), 103–8.

260. they had undergone greater spiritual growth as a result of their NDEs . . . Surbhi Khanna and Bruce Greyson, 'Near-Death Experiences and Posttraumatic Growth,' *Journal of Nervous and Mental Disease* 203(10) (2015), 749–55; Bruce Greyson and Surbhi Khanna, 'Spiritual Transformation after Near-Death Experiences,' *Spirituality in Clinical Practice* 1(1) (2014), 43–55.

262. people who had NDEs described a greater sense of well-being . . . Surbhi Khanna and Bruce Greyson, 'Near-Death Experiences and Spiritual Well-Being,' *Journal of Religion and Health* 53(6) (2014), 1605–15.

262. they also reported more daily spiritual experiences . . . Surbhi Khanna and Bruce Greyson, 'Daily Spiritual Experiences before and after Near-Death Experiences,' *Psychology of Religion and Spirituality* 6(4) (2014), 302–9.

262. people who have had NDEs report a heightened sense of purpose . . . Steven A. McLaughlin and H. Newton Malony, 'Near-Death

Experiences and Religion: A Further Investigation,' *Journal of Religion and Health* 23(2) (1984), 149–59; Cassandra Musgrave, 'The Near-Death Experience: A Study of Spiritual Transformation,' *Journal of Near-Death Studies* 15(3) (1997), 187–201; Bruce Greyson, 'Near-Death Experiences and Spirituality,' *Zygon* 41(2) (2006), 393–414; Natasha A. Tassell-Matamua and Kate L. Steadman, ' "I Feel More Spiritual": Increased Spirituality after a Near-Death Experience,' *Journal for the Study of Spirituality* 7(1) (2017), 35–49.

262. such a strong personal connection to the divine that religious observances seem unnecessary . . . Greyson, 'Near-Death Experiences and Spirituality'; Tassell-Matamua and Steadman, ' "I Feel More Spiritual" '; Kenneth Ring, *Life at Death: A Scientific Investigation of the Near-Death Experience* (New York: Coward, McCann & Geoghegan, 1980).

266. experiencers express greater compassion and concern for others . . . Russell Noyes, Peter Fenwick, Janice Miner Holden, and Sandra Rozan Christian, 'Aftereffects of Pleasurable Western Adult Near-Death Experiences,' in *The Handbook of Near-Death Experiences*, ed. by Janice Miner Holden, Bruce Greyson, and Debbie James (Santa Barbara, CA: Praeger/ABC-CLIO, 2009), 41–62.

266. Every major religion has some variant of this as one of its basic guidelines . . . Antony Flew, *A Dictionary of Philosophy* (London: Pan Books, 1979), 134; William Spooner, 'The Golden Rule,' in *Encyclopedia of Religion and Ethics, Volume 6*, ed. by James Hastings (New York: Charles Scribner's Sons, 1914), 310–12; Simon Blackburn, *Ethics* (Oxford: Oxford University Press, 2001), 101; Greg Epstein, *Good without God* (New York: HarperCollins, 2010), 115; Jeffrey Wattles, *The Golden Rule* (Oxford; Oxford University Press, 1996); Gretchen Vogel, 'The Evolution of the Golden Rule,' *Science* 303(5661) (2004), 1128–31.

267. 'If there is a God, I should live my life according to the principles of kindness . . . This quote appears on pages 196–97 of Dinty Moore, *The Accidental Buddhist* (New York: Broadway Books, 1997).

267. the Golden Rule is the result of an unconscious brain mechanism . . . Donald Pfaff and Sandra Sherman, 'Possible Legal Implications of Neural Mechanisms Underlying Ethical Behaviour,' in *Law and Neuroscience: Current Legal Issues 2010, Volume 13*, ed. by Michael Freeman (Oxford, UK: Oxford University Press, 2011), 419–32.

267. Experiencers often describe the Golden Rule not as a moral guideline . . . David Lorimer, *Whole in One* (London: Arkana, 1990).

270. spiritual lessons from NDEs as clichéd, well-trod religious platitudes . . . See, for example, Ted Goertzel, 'What Are Contact 'Experiencers' Really Experiencing?' *Skeptical Inquirer* 43(1) (2019), 57–59.

270. 'The Golden Rule is of no use to you whatever unless you realize it's your move' . . . Quoted on page 45 of Robert D. Ramsey, *School Leadership From A to Z* (Thousand Oaks, CA: Corwin Press, 2003).

271. Fran Sherwood had an NDE during emergency abdominal surgery . . . Fran described her NDE in Frances R. Sherwood, 'My Near-Death Experience,' *Vital Signs* 2(3) (1982), 7–8.

17. A NEW LIFE

273. shrapnel entered through the armhole of Steve Price's bulletproof vest . . . Steve's NDE was described in 'Fascinating Near-Death Experiences Changed Lives – Forever,' *Weekly World News*, December 19, 2000, page 35. It was also described on pages 217–27 of Barbara Harris and Lionel C. Bascom, *Full Circle* (New York: Pocket Books, 1990).

275. Mickey, who collected money for the Mafia in the 1970s, also had a profound reversal . . . Mickey's NDE was described on pages 41–44 of Charles Flynn, *After the Beyond* (Englewood Cliffs, NJ: Prentice Hall, 1986).

288. Gordon Allen was an entrepreneur, a ruthless and successful financier . . . Gordon described his NDE in *The Day I Died*, produced by Kate Broome (London: BBC Films, 2002).

18. HARD LANDINGS

294. the difficulties they were having as a result of their experiences . . . I described a few of these cases in Bruce Greyson, 'The Near-Death Experience as a Focus of Clinical Attention,' *Journal of Nervous and Mental Disease* 185(5) (1997), 327–34.

295. a five-day workshop on helping experiencers . . . Bruce Greyson and Barbara Harris, 'Clinical Approaches to the Near-Death Experiencer,' *Journal of Near-Death Studies* 6 (1987), 41–52.

296. The DSM-IV included a new category called 'Religious or Spiritual Problems' . . . American Psychiatric Association, *Diagnostic and Statistical Manual of Mental Disorders, 4th Edition* (Washington: American Psychiatric Association, 1994), 685.

296. anger, depression, and a sense of isolation can frequently arise after NDEs . . . Robert P. Turner, David Lukoff, Ruth Tiffany Barnhouse, and Francis G. Lu, 'Religious or Spiritual Problem: A Culturally Sensitive Diagnostic Category in the DSM-IV,' *Journal of Nervous and Mental Disease* 183(7) (1995), 435–44.

297. the types of help and support experiencers seek out . . . Marieta Pehlivanova, 'Support Needs and Outcomes for Near-Death Experiencers,' presented at the 2019 American Center for the Integration of Spiritually Transformative Experiences Conference, Atlanta, November 15, 2019.

298. more than fifty IANDS-affiliated groups in various cities . . . Current contact information for these groups is available at www.iands. org/groups/affiliated-groups/group-resources.html.

298. internet-based support groups through the IANDS Sharing Groups Online . . . Information about these groups is available at www.isgo.iands.org.

299. Family and friends often have difficulty understanding and adapting . . . Rozan Christian and Janice Miner Holden,' " 'Til Death Do Us Part": Marital Aftermath of One Spouse's Near-Death Experience,' *Journal of Near-Death Studies* 30(4) (2012), 207–31; Mori Insinger, 'The Impact of a Near-Death Experience on Family Relationships,' *Journal of Near-Death Studies* 9(3) (1991), 141–81.

299. marriages in which one partner has had an NDE are less well-adjusted . . . Charles P. Flynn, *After the Beyond* (Englewood Cliffs, NJ: Prentice Hall, 1986); Cherie Sutherland, *Reborn in the Light* (New York: Bantam, 1992).

19. A NEW VIEW OF REALITY

307. 'human beings are biologically "hardwired for God"' . . . This quote appears on page 178 of Andrew Newberg and Mark Robert Waldman, *Why We Believe What We Believe* (New York: Free Press, 2006).

309. meditation, a mental practice of focusing the nonphysical mind . . . Adrienne A. Taren, J. David Creswell, and Peter J. Gianaros, 'Dispositional Mindfulness Co-Varies with Smaller Amygdala and Caudate Volumes in Community Adults,' *PLOS One* 8(5) (2013), e64574.

309. meditating on their NDEs changes their physical brains . . . Mario Beauregard, Jérôme Courtemanche, and Vincent Paquette, 'Brain Activity in Near-Death Experiencers during a Meditative State,' *Resuscitation* 80(9) (2009), 1006–10.

309. psychotherapy, another nonphysical process, changes the physical brain . . . David Linden, 'How Psychotherapy Changes the

Brain – The Contribution of Functional Neuroimaging,' *Molecular Psychiatry* 11(6) (2006), 528–38; Jeffrey M. Schwartz, 'Neuroanatomical Aspects of Cognitive-Behavioural Therapy Response in Obsessive-Compulsive Disorder: An Evolving Perspective on Brain and Behaviour,' *British Journal of Psychiatry* 173 (Supplement 35) (1998), 38–44.

312. a 'benign virus' that other people can catch from experiencers . . . Kenneth Ring and Evelyn Elsaesser Valarino, *Lessons from the Light* (New York: Insight/Plenum, 1998).

312. students in a sociology class that studied NDEs felt more compassionate concern . . . Charles Flynn, *After the Beyond* (Englewood Cliffs, NJ: Prentice Hall, 1986).

312. nursing students who completed a course on NDEs had less fear of death . . . Kenneth Ring, 'The Impact of Near-Death Experiences on Persons Who Have Not Had Them: A Report of a Preliminary Study and Two Replications,' *Journal of Near-Death Studies* 13(4) (1995), 223–35.

312. students in a psychology class on NDEs had greater appreciation of life . . . Kenneth Ring, *The Omega Project* (New York: William Morrow, 1992); Ring, 'The Impact of Near-Death Experiences.'

312. Those who were exposed to the NDE information had greater appreciation for life . . . Natasha Tassell-Matamua, Nicole Lindsay, Simon Bennett, et al., 'Does Learning about Near-Death Experiences Promote Psycho-Spiritual Benefits in Those Who Have Not Had a Near-Death Experience?' *Journal of Spirituality in Mental Health* 19(2) (2017), 95–115.

313. Information about NDEs was also included in the health curriculum of a high school . . . Glenn E. Richardson, 'The Life-after-Death Phenomenon,' *Journal of School Health* 49(8) (1979), 451–53.

313. medical and nursing schools now include information about NDEs in their curricula . . . Robert D. Sheeler, 'Teaching Near Death Experiences to Medical Students,' *Journal of Near-Death Studies* 23(4) (2005), 239–47; Mary D. McEvoy, 'The Near-Death Experience: Implications for Nursing Education,' *Loss, Grief & Care* 4(1–2) (1990), 51–55.

313. more sensitive to the frequency and effects of NDEs in their patients . . . Ryan D. Foster, Debbie James, and Janice Miner Holden, 'Practical Applications of Research on Near-Death Experiences,' in *The Handbook of Near-Death Experiences*, ed. by Janice Miner Holden, Bruce Greyson, and Debbie James (Santa Barbara, CA: Praeger/ABC-CLIO, 2009), 235–58.

313. introducing information about NDEs into the treatment of suicidal clients . . . John M. McDonagh, 'Introducing Near-Death Research Findings into Psychotherapy,' *Journal of Near-Death Studies* 22(4) (2004), 269–73; Engelbert Winkler, 'The Elias Project: Using the Near-Death Experience Potential in Therapy,' *Journal of Near-Death Studies* 22(2) (2003), 79–82.

313. information about NDEs can reduce the suffering of grieving individuals . . . Mette Marianne Vinter, 'An Insight into the Afterlife? Informing Patients about Near Death Experiences,' *Professional Nurse* 10(3) (1994), 171–73; Bruce J. Horacek, 'Amazing Grace: The Healing Effects of Near-Death Experiences on Those Dying and Grieving,' *Journal of Near-Death Studies* 16(2) (1997),149–61.

20. LIFE BEFORE DEATH

319. John Wren-Lewis was poisoned by a would-be thief . . . John described his NDE in John Wren-Lewis, 'The Darkness of God: A Personal Report on Consciousness Transformation through an Encounter with Death,' *Journal of Humanistic Psychology* 28(2) (1988), 105–12.

323. a dialogue between Buddhist scholars and Western scientists on mind and matter . . . Bruce Greyson, 'Is Consciousness Produced by the Brain?' in *Cosmology and Consciousness*, ed. by Bryce Johnson (Dharamsala, India: Library of Tibetan Works and Archives, 2013), 59–87.

Index

....

Eldadah, Basil 183

electrical probe of the brain 144–7, 148, 171, 172

electroencephalogram (EEG), flat 159

elephant, fable of blind men and the 163

Elizabeth (age 28, pregnancy) 261–2

emergency room (ER), admission to 2

Emily (age 49, almost drowned) 277–8

emotions
the brain and 309
studying them 32

end of life, life review therapy at 63

endorphins, and NDEs 162–3

eternity, experience of 47–8

evidence, types of scientific evidence that are hard to verify 236–7

experiencers (those who experience an NDE)
accuracy of descriptions by 106–7
author's database of 38, 42
being referred for psychological evaluation 42
belief that some part of them will live on after death 193–4
'benign virus' spread by 312
deciding to go back to life 224, 225
effects on family members 299–300
importance of NDEs to 17
need to talk to doctors later 103
no way of knowing what they knew 200–4
no way of knowing what they saw 195–7
questionnaires filled out by 14
REM correlated with 160
as self-selected volunteers 65–6
three goals for 296

told in the afterlife to go back to life 224, 225
turning their life around 211
type of help available to 295–7
see also near-death experience (NDE)

experimental group 87

explosion, being in one 39

family counseling, help with NDEs 298

Faraday, Ann 319–20

Fitzpatrick, Róisín (age 35, brain hemorrhage) 218–19, 220

Fountain, Nathan 147–52

Franciscan nuns study 307

Freud, Sigmund 31, 236

Friel, Judy (age 24, hospitalized) 206–7, 208

frontal lobe 147

Galatians, Epistle to 266

Gandalf 228

genie-liberated-from-bottle experience 105

George (age 49, heart stopped) 243

Geraci, Joe (age 36, policeman, bleeding after surgery) 47–8, 68–9, 275, 276, 293, 314

germs, former disbelief in 133

Gibson, Arvin 89

Gina (age 24, police officer who attempted suicide) 71–5

Ginger (mother of Bobby) 300–3

Glaser, Donald 32

Glen (age 36, electrocuted) 248

Glenn, Katherine (age 27, hospitalized) 263

glutamate 162–3

gnashing of teeth in hell 212

God 78
belief in 214–15, 233–5
crying out to, to save oneself 213, 214

near-death experience – *cont.*
  long remembered 129
  reliability of memories over time
    138–9
  negative types
    behaving inappropriately after
      293
    distressing experiences 210–17,
      221
    negative effects of 284–303
    positive effects of 313
  numbers of
    frequency of after brain stops
      175–6
    frequency of during operations
      109
    frequency and normality of
      129–30, 316, 317
    normality of, and not as sign of
      mental illness 317
    one-time experience of 128
    rate of experiencing 119–20
  reality of experience 17
    experiencers' belief in reality of
      140–1
    felt as real 126–7, 153–4,
      288–90
  the term 78
    as a new concept 9, 13
    first use of term 36–7
    other terms for 140
  see also experiencers
near-death experience (NDE)
    research 37–8, 65–76,
      168–9
  as author's unpaid hobby 88–9
  considered unscientific 132
  considered a waste of time and not
    possible 86–8
  control groups in (including those
    who didn't have an NDE) 72
Nelson, Kevin 159–60
New Testament 266
Newberg, Andy 307

'No wonder, no wonder' saying while
    having an NDE 61–2
Noë, Alva 173, 174
Nome, Gregg (age 24, drowning in a
    waterfall) 45–6, 50
  life review 52–3
nonphysical things (thoughts,
    emotions)
  applying science to 15–16
  difficulty of measuring 33
  reality of 32

occipital lobe 171
odors, unique odors in an NDE 51
Ogston, Sir Alexander 103–5
  NDE of 104–5, 155
Old Testament 266
Oneness 231
online resources, help with NDEs
    298–9
opioids 123–4
organizations that help with NDEs
    298–9
Other World (afterlife, where
    experiencers go) 205–22
  blackness in 264
  called heaven 206–11
  detailed descriptions of some
    166–7, 193–4, 198–9, 212,
    273–4
  escaping from and returning to
    our world 208
  flash of lightning illuminating 252
  lights in 258
  no name or label for 208, 213,
    214
  no words to describe 68–71
  not necessarily a place 222
  remembered by experiencers 218
out-of-body-viewpoint experience
    95–103, 110–11, 126–7,
    145–7, 148, 166–7, 244,
    321
  accuracy of 106–7, 137